Transboundary Offshore Aquifers

Transboundary Offshore Aquifers

A Search for a Governance Regime

By

Renee Martin-Nagle

BRILL

LEIDEN | BOSTON

Library of Congress Control Number: 2016940157

Typeface for the Latin, Greek, and Cyrillic scripts: "Brill". See and download: brill.com/brill-typeface.

ISBN 978-90-04-32635-4 (paperback)
ISBN 978-90-04-32636-1 (e-book)

Originally published as Volume 1(2) 2016, in *International Water Law*, DOI 10.1163/23529369-12340002.

This book is printed on acid-free paper and produced in a sustainable manner.

Contents

Author Biography

Renee Martin-Nagle is a PhD Researcher in transboundary aquifer law and a lecturer in international environmental law at the University of Strathclyde in Glasgow, Scotland. In addition, she is also a Visiting Scholar at the Environmental Law Institute, Treasurer of the International Water Resources Association, and President and CEO of A Ripple Effect PLC. For more than 20 years, Renee was the chief legal officer of Airbus Americas, retiring in January 2011 as Vice President, General Counsel, Chief Compliance Officer, Head of Environmental Affairs, Corporate Secretary and a member of the board of directors. Her aviation credentials include receiving the 2014 Aviation Industry Woman of Excellence Award from the International Aviation Women's Association, chairing the American Bar Association's Forum on Air and Space Law and serving on the boards of directors of Women in Aviation International and the AeroClub of Washington. She also served on the board of directors of Omega Institute.

The birth of her first grandson in 2007 inspired Renee to dedicate her professional energies to environmental issues, and, while still working full-time for Airbus Americas, she earned an LL.M. in environmental law from George Washington University in Washington, D.C., graduating in 2010 with highest honors. In addition, her LL.M. thesis, 'Fossil Aquifers: A Common Heritage of Mankind,' received the 2010 Jamie Grodsky Prize for Environmental Scholarship. She has received a Juris Doctor from the University of Pittsburgh School of Law, a Bachelor of Arts from Saint Francis University and an Associate of Arts from Mount Aloysius College. Martin-Nagle has spoken frequently on a variety of freshwater issues, including the law of transboundary aquifers, governance of surface water and groundwater, bulk water sales, groundwater management during climate change, and interspecies equity.

Transboundary Offshore Aquifers

A Search for a Governance Regime

Renee Martin-Nagle
PhD Researcher, University of Strathclyde; Visiting Scholar, Environmental Law Institute; Treasurer, International Water Resources Association

Abstract

While the existence of freshwater offshore aquifers under continental shelves has been known for decades, none discovered thus far straddles international boundaries. In the event that an offshore aquifer shared by two or more nations is identified and targeted for development, selection of a governance regime for the aquifer will present a unique challenge, although several current legal systems could provide guidance. The UN Convention on the Law of the Sea addresses maritime natural resources in formations such as the continental shelf. Well-established governance principles for development of offshore hydrocarbon reserves would also be relevant. While the law addressing transboundary land-based aquifers is still in a nascent stage, customary international water law has evolved over centuries and could provide guidance for offshore aquifers. This monograph explores principles, and proposes a governance regime for transboundary offshore aquifers.

Keywords

continental shelf – freshwater aquifers – marine resources – off-shore aquifers – seabed hydrocarbons

Introduction

This monograph will explore the legal implications of a natural resource whose significance is just beginning to be appreciated—offshore aquifers containing fresh to brackish waters that lie within continental shelves. As the first chapter will demonstrate, fresh to slightly brackish water has been found under continental shelves along several continents, and scientists have postulated that water-laden geological formations lying hidden in seabeds around the world may be the rule rather than the exception. Thus far, since all of the

 | DOI 10.1163/9789004326361

offshore aquifers that have been discovered to date lie within the 200-mile Exclusive Economic Zone (EEZ) of a single state, domestic laws of the relevant state have governed their ownership. Hence, the states within whose continental shelves the currently identified offshore aquifers exist have either full sovereignty or full sovereign rights over the resource, depending on whether the aquifer lies within the 12-mile territorial sea or the EEZ. However, one can easily imagine discovery of an offshore aquifer in the not-so-distant future that spreads across a boundary line separating two or more states, transforming the aquifer into a transboundary resource. Should predictions of vast offshore reserves be accurate, and should those offshore reserves lie in maritime areas claimed by two or more states that desire to access and utilize the water, then conflicts could arise with respect to development rights. In a water-scarce world, clear and equitable governance regimes may prevent hostile disputes over a vital commodity.

In the event that transboundary offshore aquifers (TBOA) are discovered and their development is deemed to be practical, which governance regime and legal principles should provide guidance to the states involved? Should those states look for guidance in treaties such as the Convention on the Continental Shelf[1] and the United Nations Convention on the Law of the Sea (UNCLOS)[2]? The fact that the TBOAs are located in maritime territory would seem to argue in favor of following a set of maritime principles that were produced after decades of negotiations, culminating in UNCLOS, which now has 167 parties.[3] The International Court of Justice (ICJ) has opined on how maritime transboundary resources should be allocated, so states can look to analyses and pronouncements from a global court as well. The second chapter of this monograph will explore legal concepts of offshore natural resources arising from global maritime treaties and ICJ decisions.

1 Convention on the Continental Shelf, 29 April 1958, 15 U.S.T. 471, T.I.A.S. No. 5578, 499 U.N.T.S. 311, *available at* http://cil.nus.edu.sg/1958/1958-convention-on-the-continental-shelf/ (accessed 7 February 2016).

2 The United Nations Convention on the Law of the Sea (UNCLOS), 10 December 1982, 1833 U.N.T.S. 897, *available at* http://www.un.org/depts/los/convention_agreements/texts/unclos/unclos_e.pdf (accessed 7 February 2016).

3 As of 7 February 2016, UNCLOS has 157 signatories and 167 parties. https://treaties.un.org/pages/ViewDetailsIII.aspx?src=TREATY&mtdsg_no=XXI-6&chapter=21&Temp=mtdsg3&lang=en (accessed 7 February 2016).

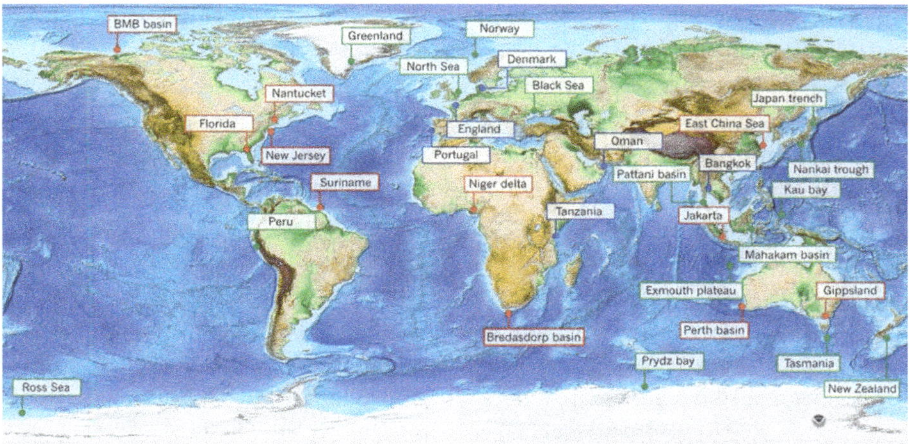

IMAGE 1 *Map of known and proposed offshore freshwater aquifers. Red boxes indicate offshore freshwater aquifers proven by observational data. Green boxes indicate offshore aquifers that may not be entirely fresh, but some fresh water can be inferred based on pore-water composition. Blue boxes indicate places where indirect on-shore evidence indicates the existence of offshore aquifers.*
REPRINTED WITH PERMISSION FROM MACMILLAN PUBLISHERS LTD: NATURE, VOL. 504, COPYRIGHT (2013).

However, a rich body of treaties, contracts and judicial and arbitral opinions has also resulted from decades of offshore hydrocarbon exploration and extraction by the oil and gas industry. States with TBOAs could therefore look for analogies in the customs and practices that have developed with respect to offshore oil and gas development since oil and gas are also fluid resources residing in seabeds. The third chapter of this monograph will therefore present a discussion of established legal concepts relating to states' rights with respect to offshore oil and gas.

Yet a third source of guidance exists in the still-evolving body of hard and soft laws pertaining to land-based surface water and transboundary freshwater aquifers (TBA). While still in a nascent stage, the treaties and customary laws regarding surface water and groundwater in TBAs could also provide insights into governance of TBOAs and will thus be discussed in the fourth chapter.

The final chapter will review pertinent concepts from all three potential sources of legal guidance for TBOAs and will attempt to provide a framework for states to follow should a TBOA suitable for development be discovered within their jurisdictions.

Part 1: Origin of Offshore Submarine Groundwater

Every now and then, a group of curious and diligent scientists produces a breakthrough in understanding by investigating an anomaly, and that is how offshore submarine aquifers were first identified in the 1970s by a team at the U.S. Geological Survey (USGS) working on Nantucket Island, Massachusetts.[4] Over the ensuing decades, scientists have studied the phenomenon more closely, as subsequent sampling has revealed similar aquifers lying under continental shelves around the world. Further, modeling has predicted that these aquifers could contain enough fresh or slightly brackish water to serve as viable reservoirs of a vital resource. To understand how these pockets of fresh water were formed and escaped detection for so long, one must delve into distant geological periods and glacial processes and explore their origins through a synergy of hydrogeology, climatology and glaciology. As will be explained below, offshore submarine aquifers can be classified into two main types: those that originated during the glacial periods and those hydraulically connected to on-shore aquifers.[5]

Discovery of Offshore Aquifers

As noted in the previous paragraph, the existence of offshore aquifers was discovered in the 1970s when a group of scientists at the USGS drilled test holes in the center of Nantucket Island to determine the thickness of a freshwater aquifer there and to analyze the geologic formations under the island down to the bedrock.[6] To their surprise, they found two additional freshwater aquifers lying beneath the known freshwater aquifer.[7] Seawater is defined as having a salinity of 33 to 35 parts per thousand (PPT),[8] while fresh water is defined as having a salinity of less than 0.5 PPT.[9] The two anomalous caches of fresh water measured 0.5 PPT from 730 feet (222 meters) to 820 feet (250 meters) and 0.5 PPT at 950 feet (286 meters), with a layer of increased salinity to 12 PPT at 820–840

4 F.A. Kohout et al, 'Fresh Ground Water Stored in Aquifers Under the Continental Shelf: Implications from a Deep Test, Nantucket Island, Massachusetts,' *Water Resources Bulletin* Vol. 13(2), pp. 373–386 (1977).

5 Ibid, p. 384; J. Groen et al, 'Fresh and Moderately Brackish Groundwaters in Coastal Plains and Continental Shelves: Past and Ongoing Natural Processes,' Proc. SWIM [Salt Water Intrusion Meeting] 16, Poland, pp. 1–7 (2000), p. 1.

6 Supra, Kohout et al, note 4, p. 384.

7 Ibid, p. 383.

8 Ibid, p. 378.

9 Ibid, pp. 381–382.

feet (250–256 meters).[10] These low salinity levels meant that the aquifers were suitable for human use. For the island, the presence of an additional supply of fresh water was probably very good news, while for the scientific community the discovery presented an interesting opportunity for further investigation.

Logically, an island off the East Coast of the United States should not be the only place in the world to house hidden troves of fresh water. Hence, the USGS findings on Nantucket Island initiated a search for more liquid treasure along the Atlantic coastline. A subsequent study found fresh water off the coast of New Jersey and concluded that the transition zone from fresh water to seawater probably reflected emplacement of fresh water earlier than the Pleistocene Epoch.[11] Almost simultaneously, the presence of offshore fresh water near Holland was also announced.[12] Soon, Groen et al. were predicting that 'fresh/brackish waters in the offshore may be the rule rather than the exception.' Since heavier seawater will eventually infiltrate the freshwater formations,[13] recommendations were made that utilization of offshore aquifers be considered for human needs sooner rather than later.[14] Subsequent exploration found fresh and moderately brackish water extending 25 kilometers from the coast of Suriname into its continental shelf, and the age of the groundwater as well as the geological conditions led the researchers to conclude that the offshore aquifer has no contemporary recharge.[15] A further study suggested that under the right conditions, ancient groundwater could extend for tens of kilometers beyond coastlines and into continental shelves.[16]

---

10 Ibid, p. 382.

11 Harold Meisler, P. Patrick Leahy & Leroy L. Knobel, 'Effect of Eustatic Sea-Level Changes on Saltwater-Freshwater in the Northern Atlantic Coastal Plain,' Water-Supply Paper 2255, *U.S. Geological Survey*, pp. 1–28 (1984), p. 26.

12 Ibid, p. 16.

13 Depending on the composition of the over-lying confining layer, seawater infiltration has been estimated to require tens of thousands of years. Tor Haakon Bakken, Fridtjof Ruden & Lars Erik Mangset, 'Submarine Groundwater: A New Concept for the Supply of Drinking Water,' *Water Resource Management*, Vol. 26, pp. 1015–1026 (2012), p. 1018.

14 Supra, Groen et al, note 5, p. 5.

15 Ibid, p. 3, *referencing* Groen, J., 'Hydrogeological Investigations in Suriname,' *The History of Earth Sciences in Suriname*. (1998); J. Groen, J. Velstrab & A.G.C.A. Meesters, 'Salinization Processes in Paleowaters in Coastal Sediments of Suriname: Evidence from ∂^{37}Cl Analysis and Diffusion Modelling,' *Journal of Hydrology*, Vol. 234, pp. 1–20 (2000), p. 18.

16 H. Kooi & J. Groen, 'Offshore Continuation of Coastal Groundwater Systems; Predictions Using Sharp-interface Approximations and Variable-density Flow Modelling,' *Journal of Hydrology*, Vol. 246, pp. 19–35 (2001), p. 31.

Attention then turned to finding palaeo-groundwater in Europe through a project called 'PALAEAUX.'[17] PALAEAUX funded a search for coastal groundwaters that were emplaced up to 50,000 years ago from the Baltic to the Canary Islands; the time frame of 50,000 years was selected since it is the upper limit for radiocarbon ^{14}C dating.[18] For the most part, the researchers discovered that the palaeo-groundwater had been in a steady state condition since sea levels in Europe stabilized 7,000 years ago,[19] and thus would not contain the pollutants that have been emitted during the industrial era.[20] While most of the studies in the PALAEAUX project focused on land-based palaeowater, several of the PALAEAUX studies found compelling data strongly suggesting the presence of fresh and brackish aquifers off the coasts of the United Kingdom, Denmark and Portugal.[21]

Glacial Origins

The story of how water came to be emplaced offshore begins millions of years ago. As most people know, the continents we have come to recognize have not always been in their current configurations. At least four times in the murky past all of the land masses of the globe were combined into super-continents.[22] The first three occurred in the pre-Cambrian Eon: Columbia, which was formed 1800–1500 million years ago (mya),[23] Rodinia, which lasted from approximately 1100 mya to 700 mya,[24] and Pannotia, which followed

17 W. M. Edmunds & C. J. Milne, eds., *Palaeowaters in Coastal Europe: Evolution of Groundwater since the Late Pleistocene* (The Geological Society of London 2001).

18 W. M. Edmunds, 'Palaeowaters in European Coastal Aquifers—The Goals and Main Conclusions of the PALAEAUX Project,' in W. M. Edmunds & C. J. Milne, eds., *Palaeowaters in Coastal Europe: Evolution of Groundwater since the Late Pleistocene*, pp. 1–16 (The Geological Society of London 2001), p. 2.

19 Ibid, pp. 2, 5.

20 Ibid, p. 2. However, drilling during the past 100 years has disturbed the natural balance of waters and initiated mixing of palaeowater and surface water that has been tainted with pollutants.

21 Ibid, p. 14; K. Hinsby et al, 'The Ribe Formation in Western Denmark—Holocene and Pleistocene Groundwaters in a Coastal Miocene Sand Aquifers,' in W. M. Edmunds & C. J. Milne, eds., *Palaeowaters in Coastal Europe: Evolution of Groundwater since the Late Pleistocene*, pp. 29–48 (The Geological Society of London 2001), pp. 42–43.

22 Scientific literature differs as to the number of super-continents that formed. *Compare* John J.W. Rogers & M. Santosh, 'Supercontinents in Earth History,' *Gondwana Research*, Vol. 6 (3), pp. 357–369 (2003) *with* J. Brendan Murphy, R. Damian Nance & Peter A. Cawood, 'Contrasting Modes of Supercontinent Formation and the conundrum of Pangea,' *Gondwana Research*, Vol. 15, pp. 408–420 (2009).

23 Ibid, Rogers & Santosh, p. 357.

24 Ibid, Rogers & Santosh.; supra, Murphy, Nance & Cawood, note 22, p. 410.

400 million years later.[25] The most recent super-continent, Pangea, came together during the Paleozoic Period approximately 250 mya,[26] and its disintegration during the Jurassic Period 200–145 mya created two super-subcontinents—Laurasia, which later formed North America, Europe and most of Asia, and Gondwana, which gave birth to South America, Africa, India, Antarctica and Australia.[27] When those super-subcontinents broke apart and reformed into their now-familiar positions,[28] new continental edges were formed, and the action of meteoric rain that washed silts and soil from the land masses to the oceans over millions of years caused geological layers to develop along those continental edges and under what are now marine continental shelves.[29]

Another series of climatic events then took its own turn in affecting the continental shelves. A succession of glacial and interglacial periods caused large volumes of snow to form massive ice sheets that advanced southward from the North Pole before melting, retreating and then repeating the process.[30] With each advance, topography changed under the tremendous weight and momentum of the glacial masses, and rocks and soil were pushed forward and then deposited, often along coastal margins, as the glaciers shrank.[31] With so much of the planet's water conscripted into ice sheets, sea levels were as much as 400 feet lower than they are now, thus exposing currently submerged continental shelves to meteoric elements such as wind and rain for approximately a million years and allowing fresh water to infiltrate the soil and rock formations.[32] The action of glacial forces created overlying layers of compacted, over-consolidated clay[33] and other relatively impermeable substances that sealed the fresh water accumulating on exposed continental shelves into

25 Ibid, Murphy, Nance & Cawood.

26 Ibid, p. 408; Supra, Rogers & Santosh, note 22, p, 366.

27 Kangshan Mao et al, 'Distribution of Living Cupressaceae Reflects the Breakup of Pangea,' *Proceedings of the National Academy of Science*, Vol. 109 (20), pp. 7793–7798 (2012), p. 7793.

28 Steven Dutch, 'Plate Tectonics and Earth History,' available at https://www.uwgb.edu/dutchs/EarthSC102Notes/102PTEarthHist.htm (accessed 28 October 2015).

29 Supra, Harold Meisler, P. Patrick Leahy & Leroy L. Knobel, note 11, p. 5; Denis Cohen et al, 'Origin and Extent of Fresh Paleowaters on the Atlantic Continental Shelf, USA,' *GROUND WATER*, Vol. 48(1), pp. 143–158 (2010), p. 144.

30 Supra, Kohout et al, note 4, pp. 375, 377.

31 Mark Person et al, 'Pleistocene Hydrogeology of the Atlantic Continental Shelf, New England,' *GSA Bulletin*, Vol. 115, (11), pp. 1324–1343 (2003), p. 1339.

32 Supra, Kohout et al, note 4, 384.

33 Mark Person et al, 'Pleistocene Hydrology of North America: The Role of Ice Sheets in Reorganizing Groundwater Flow Systems,' *Reviews of Geophysics*, Vol. 45, Paper number 2006RG000206, pp. 1–28 (2007), p. 18.

aquifer formations.[34] The water was thus protected from salinization during the inter-glacial sea level rises that flooded and then covered those shelves. Much of the pre-historic water emplacement, which could extend tens to hundreds of kilometers from shore, occurred during the Pleistocene Epoch that existed between 1.8 mya and 11,700 years ago,[35] causing these waters to be dubbed "palaeo-water."[36] The last glacial maximum, which marked the point of highest concentration of water in glaciers and therefore the lowest point for sea levels, occurred between 26,500 and 19,000 years ago.[37] Studies have shown that sea levels have been lower than they are today for 90% of the past 120,000 years, and have not been as high as they are now in the past 100,000 years.[38] Clearly, the confining layers protecting the offshore submarine groundwater bodies have been effective for many years, although with time saline water will infiltrate these aquifers due to its weight and density, which are higher than those of fresh water.[39]

Around the same time that the results of the PALAEAUX project were being released, other studies were published indicating that offshore groundwater did not originate solely from the exposure of continental shelves to meteoric conditions during the last glacial maximum, but may also have been the result of interactions between glaciers and aquifer systems. For example, the weight and forward movement of glaciers created bulges at their terminal edges, thus pushing coastal land masses further above sea level and exposing even more of the continental edges to meteoric elements.[40] Further, a group of scientists conducting additional analyses of the Atlantic coast suggested that the enormous heights at the heads of the glaciers produced high recharge rates as the glaciers melted, facilitating infusion of fresh water into the continental shelves.[41] With the combination of exposed continental shelves, glacially forced recharge, permeable rock formations and a 45-meter thick confining

34 Supra, J. Groen, J. Velstrab & A.G.C.A. Meesters, note 15, p. 2.

35 Kim Ann Zimmerman, 'Pleistocene Epoch: Facts about the Last Ice Age,' livescience, http://www.livescience.com/40311-pleistocene-epoch.html (accessed 26 December 2015).

36 Supra, Groen et al, note 5, p. 1; supra, Edmunds, note 18, p. 2.

37 Peter U. Clark et al, 'The Last Glacial Maximum,' *Science*, Vol. 325, pp. 710–714 (2009), p. 711.

38 Supra, Edmunds, note 18, p. 5.

39 Supra, Bakken, Ruden & Mangset, note 13, p. 1018.

40 Kurt Lambeck, John Chappell, 'Sea Level Change Through the Last Glacial Cycle,' *Science*, Vol. 292, pp. 679–686 (2001), p. 681.

41 Mark Person et al, 'Pleistocene Hydrogeology of the Atlantic Continental Shelf, New England,' *GSA Bulletin*, Vol. 115 (11), pp. 1324–1343 (2003), p. 1327.

layer, Atlantic coast aquifers were predicted to extend approximately 80 kilometers beyond western limit of the continental shelf and approximately 30 kilometers beyond the mean sea level during the Pleistocene.[42] Meanwhile, core drilling revealed yet more Pleistocene palaeowater lying 10 nautical miles (18.5 kilometers) off the coast of Lima, Peru.[43]

Other Origins of Offshore Aquifers

Entrapment of meteoric fresh water during glacial periods and glacial infusions of water into the ground are not the only methods of emplacing the fresh water found in continental shelves. Submarine bodies of fresh water can also be hydraulically connected to land-based aquifers, with water flowing from shore to shelf.[44] Indeed, evidence of groundwater movement from land into submarine areas has been found in the continental shelves of every continent.[45] Most estimates state that the global volume of submarine groundwater discharge is between 0.01% and 10% of surface water runoff into the oceans.[46] One study concluded that the volume of submarine fresh groundwater flow along the South Atlantic Bight of the southeast U.S. comprised up to 40% of river water runoff into that coastal area.[47]

---

42 Ibid, p. 1337.

43 Cornelia Kriete, Axel Suckow & Bodo Harazim, Pleistocene Meteoric Pore Water in Dated Marine Sediment Cores off Callao, Peru,' *Estuarine, Coastal and Shelf Science*, Vol. 59, pp. 499–510 (2004), p. 499.

44 John C. Hathaway et al, 'U.S. Geological Survey Core Drilling on the Atlantic Shelf,' *Science*. Vol. 206 (4418), pp. 515–527 (1979), p. 523; Thomas M. Church, 'An Underground Route for the Water Cycle,' *Nature*, Vol. 380, pp. 579–580 (1996), pp. 579–580.

45 Hugues Faure, Robert C. Walter & Douglas R. Grant, 'The Coastal Oasis: Ice Age Springs on Emerged Continental Shelves,' *Global and Planetary Change* Vol. 33, pp. 47–56 (2002), p. 52.

46 Thomas M. Church, 'An Underground Route for the Water Cycle,' *Nature*, Vol. 380, pp. 579–580 (1996), p. 579.

47 Willard S. Moore, 'Large Groundwater Inputs to Coastal Waters Revealed by ^{226}Ra Enrichments,' *Nature* Vol. 380, pp. 612–14 (1996), p. 612. Moore's article resulted in an exchange of letters that were published in *Nature* magazine, with Paul Younger of the University of New Castle deeming the estimate of 40% to be 'indeed surprising.' *See* 'Scientific Correspondence: Submarine Groundwater Discharge,' *Nature* Vol. 382, pp. 121–122 (1996). Thomas M. Church also called the 40% figure 'a surprisingly large percentage.' Supra, Church, note 46, p. 579.

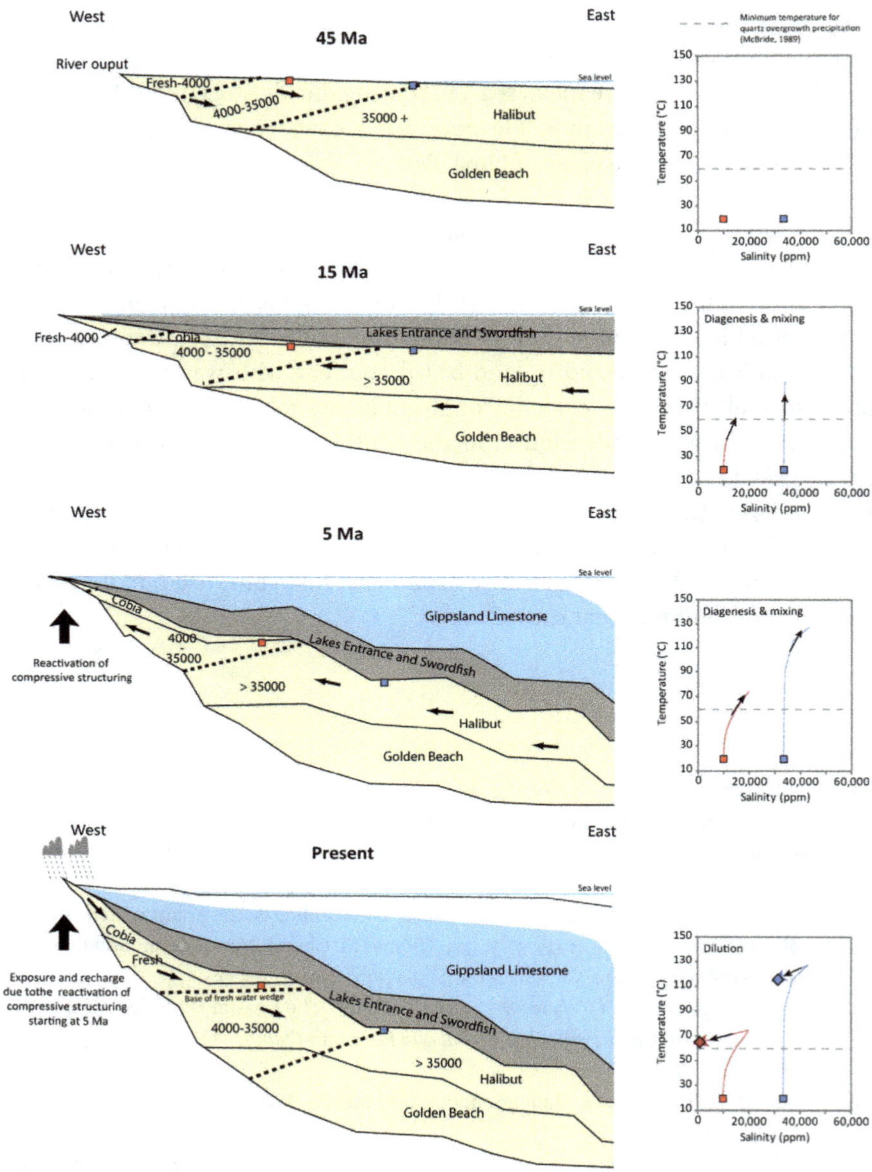

IMAGE 2 *Effect of sea level rise on continental shelf in Gippsland Basin, Australia.*
SOURCE: *GEOFLUIDS*, 2014. © 2014 JOHN WILEY & SONS LTD. REPRINTED WITH PERMISSION.

Recent Discoveries and Possible Uses

With interest in submarine groundwater growing, further tests were conducted in other places around the globe. Both Pleistocene-era embedded fresh groundwater and offshore groundwater discharge derived from current ice sheet melting were discovered off the southeastern coast of Greenland,[48] testing revealed brackish water lying 20 kilometers off the southeast shore of Australia in Gippsland Basin,[49] Indonesia was found to have offshore reserves,[50] and drilling confirmed that a coastal freshwater aquifer system in Tanzania extended beyond the coast into the Indian Ocean.[51] A team of scientists further postulated that fresh submarine groundwater could be present around volcanic islands that are part of continental shelves.[52] As testing and modeling implied ever-greater supplies of submarine groundwater, techniques for treatment of recovered brackish submarine groundwater were proposed, including sand filtration, reverse osmosis, air stripping, re-mineralization and chlorination.[53]

With the presence of offshore bodies of fresh water being confirmed through drilling in multiple locations and with extensive and sophisticated modeling indicating pervasive existence around the world, more scientists began to suggest that fresh palaeo-groundwater, untainted by industrial age pollution, could be considered as viable sources of drinking water.[54] In 2010, one group

48 Whitney DeFoor et al, 'Ice Sheet–derived Submarine Groundwater Discharge on Greenland's Continental Shelf,' *Water Resources Research*, Vol. 47(W07549), pp. 1–14 (2011), p. 1.

49 Sunil Varma & Karsten Michael, 'Impact of Multi-purpose Aquifer Utilisation on a Variable-density Groundwater Flow System in the Gippsland Basin, Australia,' *Hydrogeology Journal*, Vol. 20, pp. 119–134 (2011), p. 132; J. Bourdet, R. Kempton & K. Michael, 'Palaeo-formation Water Evolution in the Latrobe Aquifer, Gippsland Basin, South-eastern Australia Continental Shelf,' *Geofluids*, Vol. 14, pp. 1–24 (2014), p. 2.

50 Supra, Groen, Velstrab & Meesters, note 15, p. 1, *citing* Maathuis H. Mak W. & Adi S. 'Hydrogeology of the Coastal Region of Jakarta, Indonesia' in: *Groundwater: Past Achievements and Future Challenges* (ed. Sililo O) (Proceedings of the International Association of Hydrogeologists Congress on Groundwater 2000), pp. 209–213.

51 F. Ruden, 'The Discovery of a Regional Neogene Aquifer in Coastal Tanzania,' *Coastal Aquifers: Challenges and Solutions*, Vol 1, pp. 363–372 (Instituto Geologico y Minero Espana, Madrid 2007), p. 365.

52 Yukiko Kusano et al, 'Occurrence of Old Groundwater in a Volcanic Island on a Continental Shelf; an Example from Nakano-shima Island, Oki-Dozen, Japan,' *Journal of Hydrology*, Vol. 511, pp. 295–309 (2014), p. 308.

53 Supra, Bakken, Ruden & Mangset, note 13, p. 1020.

54 Supra, Kooi & Groen, note 16, p. 32; E. Custodio, W.M. Edmunds & Y. Travi, 'Management of Coastal Palaeowaters,' in W. M. Edmunds & C. J. Milne, eds., *Palaeowaters in Coastal*

estimated the total volume of Pleistocene-era fresh (<1 PPT saline) water lying in the continental shelf off the northeast coast of the United States to be in the range of 1300 km^3; the U.S. Atlantic eastern seaboard generally could contain 10^4 km^3 of fresh water, and global continental shelves could contain a massive 3×10^5 km^3.[55]

Of course, for a resource to be viable for utilization, the cost of accessing, extracting, treating and transporting it must not be prohibitively high. One study estimated that, when compared with desalination of seawater, the process of recovering submarine groundwater would be only 10% of the total cost of producing potable water through desalination.[56] However, Cohen et al cautioned that offshore aquifers are a non-renewable resource whose exploitation should be carefully considered and in addition outlined some of the technical barriers to development of offshore aquifers:

> Development of these resources would be expensive and would require offshore drilling and construction of pipelines to transfer water inland, and/or the use of modern land-based petroleum drill rigs capable of drilling large-scale (greater than 10 km) horizontal wells to access the offshore resource. Marine seismic surveys would need to be conducted to ensure that drilling hazards are avoided, such as natural gas deposits hosted in structural traps. In addition, extensive hydrologic and geophysical characterization would be necessary to map the distribution of groundwater salinity offshore and identify viable drilling targets.[57]

Europe: Evolution of Groundwater since the Late Pleistocene, pp. 313–327 (The Geological Society of London 2001), p. 315.

55 Supra, Cohen et al, note 29, p. 143. The authors caution that conditions different from the assumptions used in modeling could affect the estimate by a factor of ±2. P. 154. To put the volumes in perspective, according to the U.S. Geological Survey, the water estimated by Cohen et al to lie under the Atlantic seaboard—10^4 km^3—is equivalent to the water in all of the world's rivers. In addition, the volume estimated to lie under the global continental shelves—3×10^5 km^3—is equivalent to more than three times the volume of water in surface lakes and rivers, or all of the ground ice and permafrost on the planet. 'How Much Water Is There On, In, and Above the Earth?,' The USGS Water Science School, http://water.usgs.gov/edu/earthhowmuch.html (accessed on 5 March 2016).

56 Supra, Bakken, Ruden & Mangset, note 13, p. 1022.

57 Supra, Cohen et al, note 29, p. 155.

Since offshore aquifers have been discovered while drilling for hydrocarbons[58] and since the offshore hydrocarbon industry has decades of technical experience in extraction, use of techniques honed by that industry to access submarine groundwater may result in more efficient recovery and lower costs.[59] The true costs and efficiency of recovery and utilization may not be fully known until development begins.

Vast Meteoric Groundwater Reserves

In a logical progression of inquiry following the growing body of evidence that offshore submarine groundwater was present in many places around the world, Dr. Vincent E.A. Post and other scientists who had been researching and publishing on the phenomenon for years—Jacobus Groen, Henk Kooi, Mark Person, Shemin Ge and W. Mike Edmunds—produced an article in the December 2013 issue of *Nature* magazine that summarized and expanded on the concept of offshore submarine groundwaters.[60] Unlike previous publications, which were reviewed mainly in scientific communities, the *Nature* article attracted interest from a wide cross-section of mainstream news outlets that seized on the possibility that the global water crisis had been solved by the discovery of vast quantities of fresh water. According to Altmetric, an online service that tracks readership of scientific publications, in the days following its release the 2013 *Nature* article was mentioned in nine blogs, 137 tweets, 10 Facebook pages and 30 news outlets,[61] including United Press International (UPI),[62]

---

58 Supra, Faure, Walter & Grant, note 45, p. 53.

59 Tor Haakon Bakken & Lars Erik Mangset, 'Sub-marine Groundwater for the Supply of Drinking Water. A Review of the Hydro-geological Potential and its Technical and Economical Feasibility,' *Geophysical Research Abstracts*, Vol. 12, EGU2010–3309 (2010); supra, Bakken, Ruden & Mangset, note 13, p. 1020.

60 Vincent E.A. Post et al, 'Offshore Fresh Groundwater Reserves as a Global Phenomenon,' *Nature*, Vol. 504, pp. 71–78 (2013).

61 'Offshore Fresh Groundwater Reserves as a Global Phenomenon: Overview of Attention for Article Published in Nature,' December 2013, Altmetric, http://www.altmetric.com/details/1955436 (accessed 7 February 2016).

62 Ananth Baliga, 'Vast Freshwater Reserves Discovered under Ocean Floor,' UPI, http://www.upi.com/Science_News/Blog/2013/12/09/Vast-freshwater-reserves-discovered-under-ocean-floor/6711386609987 (accessed 7 February 2016).

Time,[63] *Science World Report*,[64] *The Huffington Post*,[65] *International Business Times*,[66] *Science Daily*,[67] *The Times of India*,[68] *Medical News Today*[69] and *Business Insider*.[70]

What did the December 2013 *Nature* article say that was remarkable enough to attract worldwide attention to the topic of offshore submarine groundwater? First, the article summarized all the previous work in understandable terms while providing a visually appealing map of known and projected reserves.[71] Secondly, it coined the term 'vast meteoric groundwater reserves' (VMGR) and estimated that the volume of fresh and brackish water[72] in offshore VMGR may be twice the volume of water that has been extracted from land-based aquifers since 1900 (4,500 km^3).[73] Lastly, the authors repeated

63 Denver Nicks, 'Immense Freshwater Reserves Discovered Beneath Ocean Floor,' *Time*, http://science.time.com/2013/12/09/immense-freshwater-reserves-discovered-beneath-ocean-floor (accessed 7 February 2016).

64 Benita Matilda, 'Scientists Discover Untapped Freshwater Reserves Beneath the Oceans,' *Science World Report*, http://www.scienceworldreport.com/articles/11464/20131209/scientits-disocver-untapped-freshwater-reserves-beneath-the-oceans.htm (accessed 7 February 2016).

65 Emily Thomas, 'Massive Freshwater Supply Found Trapped Under the Ocean,' *Huffpost Green*, http://www.huffingtonpost.com/2013/12/05/freshwater-found-under-ocean-_n_4392821.html (accessed 7 February 2016).

66 Philip Ross, 'Freshwater Reserves Under Ocean Floors "100 Times Greater" than What We've Used Since 1900,' *International Business Times*, http://www.ibtimes.com/freshwater-reserves-under-ocean-floors-100-times-greater-what-weve-used-1900-1499766 (accessed 7 February 2016).

67 '*Vast Freshwater Reserves Found Beneath the Oceans,' Science Daily*, http://www.sciencedaily.com/releases/2013/12/131208085304.htm?utm_source=feedburner&utm_medium=feed&utm_campaign=Feed%3A+sciencedaily+%28ScienceDaily%3A+Latest+Science+News%29 (accessed 7 February 2016).

68 'Scientists Discover Vast Undersea Freshwater Reserves,' *The Times of India*, http://timesofindia.indiatimes.com/home/environment/the-good-earth/Scientists-discover-vast-undersea-freshwater-reserves/articleshow/26890776.cms (accessed 7 February 2016).

69 '*Scientists Find Vast New Freshwater Sources under the Sea,' Medical News Today*, http://www.medicalnewstoday.com/releases/269771.php (accessed 7 February 2016).

70 'Scientists Discover Massive Freshwater Reserves Underneath the Ocean Floor,' *Business Insider*, http://www.businessinsider.com/scientists-discovered-massive-freshwater-reserves-underneath-the-ocean-floor-2013-12?IR=T (accessed 7 February 2016).

71 Supra, Post et al, note 60, p. 72.

72 For the purposes of the analysis in the *Nature* article, VMGR was limited to those reserves having a minimum horizontal extent of more than 10 kilometers in length, and "brackish" was defined as water having one-third the salinity of seawater. Ibid, p. 71.

73 Ibid, p. 75.

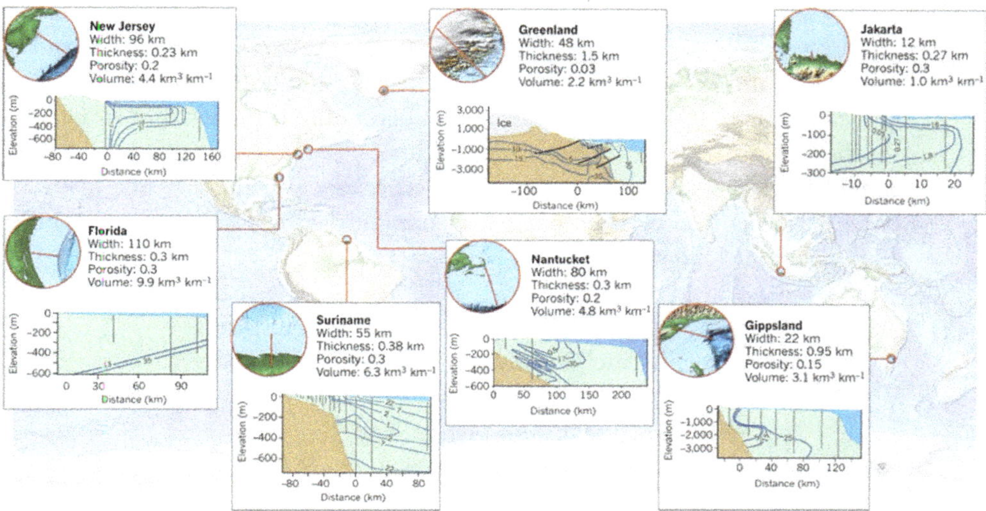

IMAGE 3 *Inferred metrics of known offshore freshwater aquifers. In the cross-sections, the blue contour lines indicate total dissolved solid (TDS) concentrations (g l⁻¹); distance (km) and elevation (m) relative to mean sea level are indicated along the horizontal and vertical axis, respectively; vertical grey lines indicate well locations where salinity is inferred from water samples and borehole logs; crystalline bedrock or low-permeability sedimentary rocks containing salt groundwater are shown in brown; the black, sub-horizontal lines denote faults; undifferentiated continental shelf sediments are in pale green; and seawater is pale blue. Within the Nantucket and Greenland cross sections, salinity contours are based on numerical model results and well data. The inferred widths, lengths and volumes per kilometer of coastline pertain to the groundwater with a TDS concentration of less than 10 g l⁻¹.*
REPRINTED WITH PERMISSION FROM MACMILLAN PUBLISHERS LTD: *NATURE* VOL. 504, COPYRIGHT (2013).

suggestions from prior publications that these offshore VMGR could be utilized to supplement drinking water supplies in drought-prone areas at costs below those of desalination.[74]

Assuming that testing continues to support projections of vast quantities of fresh to brackish water in offshore submarine aquifers, and assuming that exploration, extraction, treatment and transportation are not prohibitively expensive, then those aquifers may be viewed as viable sources of water reserves as other, land-based reserves become exhausted and/or polluted. Once development of offshore fresh water reserves begins, issues of sovereignty and governance will have to be resolved. All of the offshore groundwater bodies

74 Ibid, p. 76.

discovered thus far have been within the territory of only one nation—the United States, Suriname, Peru, Indonesia, Tanzania, Greenland and Australia—and would therefore be considered as domestic aquifers. However, one can easily imagine offshore submarine groundwater fields that straddle the boundary between two or more nations. In that case, on what basis should the interested nations determine governance regimes? Three of the potential influences on governance of offshore submarine groundwater will be explored in the following pages.

Part 2: Legal Principles Governing Maritime Resources

The Convention on the Continental Shelf

Any analysis of the governance of natural resources of the continental shelves must begin with the Truman Proclamation of 1945, through which the United States unilaterally declared the riches of its continental shelf to be within its domain and thus available for its sole exploitation.[75] The preamble of the Truman Proclamation and its accompanying press release give insights into the reasons and justification for such a bold action. The press release states that '[p]etroleum geologists believe that portions of the continental shelf beyond the three-mile limit contain valuable oil deposits,' and the preamble reads as follows:

> *Whereas* the Government of the United States of America, aware of the long range world-wide need for new sources of petroleum and other minerals, holds the view that efforts to discover and make available new supplies of these resources should be encouraged; and
> *Whereas* its competent experts are of the opinion that such resources underlie many parts of the continental shelf off the coasts of the United States of America, and that with modern technological progress their utilization is already practicable or will become so at an early date; and

75 Presidential Proclamation 2667, 28 September 1945: 'Natural Resources of the Subsoil and Sea Bed of the Continental Shelf,' *10 Federal Register* 12303 (1945), 59 Stat 884. *Also available at* http://www.presidency.ucsb.edu/ws/?pid=12332 (accessed 12 November 2015). Some scholars point to a 1942 treaty between Venezuela and the United Kingdom (acting for Trinidad) as the beginning of sovereign claims over the continental shelf. *See* A. L. Morgan, 'The New Law of the Sea: Rethinking the Implications for Sovereign Jurisdiction and Freedom of Action,' 27 *Ocean Development & International Law*, Vol. 27, pp. 5–29, 12–13 (1996).

> *Whereas* recognized jurisdiction over these resources is required in the interest of their conservation and prudent utilization when and as development is undertaken; and
> *Whereas* it is the view of the Government of the United States that the exercise of jurisdiction over the natural resources of the subsoil and sea bed of the continental shelf by the contiguous nation is reasonable and just, since the effectiveness of measures to utilize or conserve these resources would be contingent upon cooperation and protection from the shore, since the continental shelf may be regarded as an extension of the land-mass of the coastal nation and thus naturally appurtenant to it, since these resources frequently form a seaward extension of a pool or deposit lying with the territory, and since self-protection compels the coastal nation to keep close watch over activities off its shores which are of the nature necessary for utilization of these resources[76]

During World War II, the necessity of having access to adequate supplies of petroleum became clear. Although the old order was gradually coming to an end,[77] states had historically agreed that their maritime jurisdictions extended only three nautical miles from shore—as far as a cannon ball could be shot.[78] With the flourish of a pen on 28 September 28 1945, Harry S. Truman, 33rd President of the United States, changed the accepted legal order. Less than six months after taking office following the death of President Franklin Roosevelt and only six weeks after Japan's surrender ended the world war, Truman gave the U.S. the promise of petroleum security.

The Truman Proclamation consisted of only three sentences, but each sentence represented an important legal concept.

> Having concern for the urgency of conserving and prudently utilizing its natural resources, the Government of the United States regards the natural resources of the subsoil and sea bed of the continental shelf beneath the high seas but contiguous to the coasts of the United States as appertaining to the United States, subject to its jurisdiction and control. In cases where the continental shelf extends to the shores of another State, or is shared with an adjacent State, the boundary shall be determined by the United States and the State concerned in accordance with equitable

---

76 An excerpt of the press release can be found at http://www.presidency.ucsb.edu/ws/?pid=12332 (accessed 12 November 2015).

77 Barry Buzan, *Seabed Politics* (Praeger Publishers 1976), pp. 4–6.

78 Ibid, p. 2; supra, Morgan, note 75, p. 7.

> principles. The character as high seas of the waters above the continental shelf and the right to their free and unimpeded navigation are in no way thus affected.

First, the Truman Proclamation did not claim ownership or sovereignty over the continental shelf or the water column, but merely stated that the 'natural resources of the subsoil and the seabed of the continental shelf' were 'appertaining to the United States' and 'subject to its jurisdiction and control.' Secondly, the proclamation states that boundary delimitations will be determined 'in accordance with equitable principles' as opposed to a formula of equidistance between the shores of the concerned states. Lastly, the water column above the continental shelf was characterized as remaining part of the high seas, preserving freedom of navigation. The Truman Proclamation thus pertained only to natural resources of the seabed, which opened the door for exploration and development of offshore oil reserves. Fisheries were the subject of another proclamation issued on the same day.[79]

In short order, other states followed the Truman Proclamation with their own proclamations.[80] Argentina, Chile, Costa Rica, Honduras and Peru claimed sovereignty over not only the seabed of the continental shelf but also over the superjacent waters,[81] and Australia, Bahrain, Brazil, the Dominican Republic, Guatemala, Iran, Israel, Kuwait, Nicaragua, Pakistan, the Philippines, Saudi Arabia, the Trucial Sheikdoms, Venezuela and the United Kingdom also claimed varying rights over their own continental shelves.[82]

With a patchwork quilt of domestic laws declaring rights over the continental shelves and their resources being enacted around the world, the United Nations International Law Commission (UNILC) at its very first session in 1949 placed the regime of the high seas and the regime of the territorial seas on an agenda of topics to be addressed by the new body, giving the regime of the high seas priority.[83] That same year, the General Assembly passed a resolution

---

79 Presidential Proclamation 2668, 28 September 1945: 'Coastal Fisheries in Certain Areas of the High Seas,' *10 Federal Register* 12304 (*1945*), 59 Stat. 885.

80 F.V.W. Penick, 'The Legal Character of the Right to Explore and Exploit the Natural Resources of the Continental Shelf,' *San Diego Law Review*, Vol. 22, pp. 765–778 (1985), p. 769.

81 Supra, Morgan, note 75, p. 11, *citing* William. Bishop Jr, *International Law, Cases and Materials, 3rd ed.*, pp. 640–644 (Boston: Little, Brown, and Co., 1962).

82 Ibid, p. 13.

83 'Report Of The International Law Commission to the General Assembly: Law of the Sea,' Yearbook of the International Law Commission, Vol. II, *United Nations International Law Commission*, pp. 254–301(1956), p. 254, para. 7 ('1956 UNILC Yearbook'), *available at* http://legal.un.org/docs/?path=../ilc/publications/yearbooks/english/ilc_1956_

during its fourth session recommending that the regime of territorial seas also be given priority,[84] and by 1950 J.P.A. Francois, the special rapporteur who was selected to coordinate the complex task of developing a legal regime for both the high seas and the territorial seas, submitted a report to be discussed by the states, including suggestions on the treatment of the continental shelf.[85] In the general debate during the meeting with the UNILC, Francois expressed concern that the organizations that had addressed the continental shelf to date, 'such as the International Law Association, the Institut de Droit International, and so forth, were doing so from a purely scientific angle or from the viewpoint of the big oil companies.'[86] At the same time, a report from the UNILC Secretariat cited international water law in commenting favorably on the practice of protecting the unity of deposits under the continental shelf,[87] which was becoming the standard for domestic oil and gas development.[88] Thus, the United Nations considered governance of the sea and the continental shelf to be so important that some of its earliest activities were focused on the topic,

v2.pdf&lang=EFS (last accessed 14 March 2016); 'Report to the General Assembly,' Yearbook of the International Law Commission, Vol. I, *United Nations International Law Commission,* (1949), p. 281, para. 19, *available at* http://legal.un.org/docs/?path=../ilc/documentation/english/reports/a_cn4_13.pdf&lang=E (last accessed 14 March 2016).

84 United Nations General Assembly Resolution 374 (IV) (1949), *available at* http://daccess-dds-ny.un.org/doc/RESOLUTION/GEN/NR0/051/93/IMG/NR005193.pdf?OpenElement (last accessed 15 November 2015).

85 'Regime of the High Seas: Rapport de J. P. A. François, Rapporteur Spécial' Yearbook of the International Law Commission, Vol. II, *United Nations International Law Commission,* pp. 36–52 (1950) (in French) ('1950 ILC Yearbook, Vol. II'), *available at* http://legal.un.org/docs/?path=../ilc/publications/yearbooks/english/ilc_1950_v2.pdf&lang=E (last accessed 14 March 2016).

86 'Regime of the High Seas: Report by Mr. François,' Yearbook of the International Law Commission, Vol. I, *United Nations International Law Commission,* pp. 179–187 (1950) p. 179, para. 1e, *available at* http://legal.un.org/docs/?path=../ilc/publications/yearbooks/english/ilc_1950_v1.pdf&lang=E (last accessed 14 March 2016).

87 Supra, 1950 ILC Yearbook, Vol. II, note 85, p. 112, para, 337. The original French text reads as follows: Le même gisement se continuera donc très fréquemment de part et d'autre de la ligne envisagée. Or, l'exploitation d'un gisement en un lieu donné réagira d'une manière certaine, au moins en matière de substances pétrolières, sur les autres parties du gisement. Le principe qu'il ne faut pas perdre de vue et dont la pratique obligera à tenir compte est celui de l'unité du gisement. Le droit international consacre déjà cette idée en matière de droit fluvial; elle s'impose d'autant plus ici que le gisement, loin de se renouveler comme le fleuve, se détruit par l'usage.

88 William T. Onorato, 'Apportionment of an International Common Petroleum Deposit,' *International & Comparative Law Quarterly,* Vol. 17, pp. 85–102 (1968), pp. 89, 92.

and its inquiry drew both concern over and inspiration from the offshore oil and gas industry.

In 1956, the UNILC issued a final report that included draft articles on what was to become the law of the sea. The draft articles were divided into two main parts according to the original mandate from the General Assembly: the territorial sea and the high seas. The part of the report on the high seas was further divided into three sections: the general regime of the high seas, the contiguous zone and the continental shelf.[89]

The section on the continental shelf consisted of seven articles, and the concepts expressed therein provided guidance on later suggestions for applicable legal regimes. First, the continental shelf was defined as consisting of the seabed and subsoil—not the superjacent water column—outside the territorial sea to a depth of 200 meters, or beyond if the water allows for exploitation of natural resources. The term 'territorial sea' was not defined in the draft articles pertaining to the continental shelf, but draft articles addressing the territorial sea provided that the UNILC 'considers that international law does not permit an extension of the territorial sea beyond twelve miles.'[90] Further, the draft articles addressing the territorial sea pronounced that states would have full sovereignty over the airspace, seabed and subsoil in the area of the territorial sea.[91] Second, coastal states were given sovereign rights, but not sovereignty, for exploration and exploitation outside of the territorial sea. Third, rights of other states to freedom of navigation, fishing, conservation of living resources and laying of submarine cables were to be protected. Fourth, states could erect structures necessary to explore and exploit their natural resources as well as establish safety zones around the structures, but those installations would not be given the status of islands. Finally, the draft articles provided that boundary delimitations would be negotiated between and among affected states, but in the absence of agreement and, unless 'special circumstances' justified another outcome, the boundary between states with opposite coasts would be the median line between them, and the boundary between states with adjacent coasts would be an equidistant line.[92]

Granting states full sovereignty over the entire continental shelf outside the territorial sea was viewed as an unacceptable infringement on freedom of the superjacent sea and the air above it. Therefore, the UNILC proposed that

89 Supra, 1956 UNILC Yearbook, note 83, p. 255, para. 23.

90 Ibid, p. 256.

91 Ibid.

92 Ibid, p. 264.

states have sovereign rights over the natural resources of the continental shelf, thus granting states exclusive right to exploit those natural resources, and prohibiting any other state from exploring or exploiting them without receiving permission from the state possessing sovereign rights. In addition, the UNILC retained the term 'natural resources' instead of the more limited term 'mineral resources.'[93]

In 1957, a year after the UNILC issued its 1956 report that included the draft articles on the law of the sea, the United Nations General Assembly passed a resolution supporting the UNILC's recommendation that a conference be convened to conclude 'one or more international conventions or such other instruments' to embody the work of the UNILC.[94] A year later, the first United Nations Conference on the Law of the Sea took place in Geneva from 24 February to 27 April 1958.[95] The draft articles produced in 1956 formed the basis for the discussions at what was called the Geneva Conference on the Law of the Sea,[96] and the 86 states attending the conference[97] formed five committees, the fourth of which focused its efforts on the continental shelf.[98] During the course of its 42 meetings,[99] Committee IV made several changes to the initial draft articles, such as including the continental shelves of islands within the definition of continental shelf[100] and adding a provision allowing coastal states to exploit subsea resources through tunneling irrespective of the depth of the water.[101] For the most part, the concepts and wording of the draft articles produced by the UNILC in 1956 were retained, and the plenary conference decided to adopt the articles in the form of a stand-alone convention[102] that was signed by

93 Ibid, p. 297.

94 United Nations General Assembly Resolution 1105 (XI) (1957), *available at* http://legal.un.org/diplomaticconferences/lawofthesea-1958/docs/english/vol_II/5_RESOLUTION_Plenary_Mtgs_vol_II_e.pdf (last accessed 15 November 2015).

95 Marjorie M. Whiteman, 'Conference on the Law of the Sea: Convention on the Continental Shelf,' *American Journal of International Law*, Vol. 52, pp. 629–659, 629 (1958).

96 Ibid, p. 630.

97 Ibid, p. 629.

98 Ibid, p. 631.

99 Ibid.

100 Ibid, p. 634.

101 Ibid, p. 654.

102 Ibid, pp. 632, 656–657.

17 of the states before they left Geneva in April 1958.[103] The Convention on the Continental Shelf[104] entered into force in 1964.[105]

In total, the 1958 Geneva Conference on the Law of the Sea, an effort that has been dubbed UNCLOS I, produced four conventions and an optional protocol: the Convention on the Territorial Sea and the Contiguous Zone; the Convention on the High Seas; the Convention on Fishing and Conservation of the Living Resources of the High Seas; the Convention on the Continental Shelf; and the Optional Protocol of Signature concerning the Compulsory Settlement of Disputes.[106] By fragmenting the burgeoning law of the sea into parts, the 1958 Geneva Convention hoped to gain acceptance of as many provisions as possible and avoid broad reservations that would undermine the effort.[107] One main issue that the conference attendees were unable to resolve concerned the breadth of the territorial sea. Thus, while the Convention on the Continental Shelf defines the continental shelf with reference to the territorial sea,[108] the Convention on the Territorial Sea and the Contiguous Zone[109] does not indicate the limits of the territorial sea. A Second United Nations Conference on the Law of the Sea was convened, once again in Geneva, from 16 March to 26 April 1960, but failed to produce an agreement on the breadth of the territorial sea.[110] By the time that the current United Nations Convention on the Law of the Sea (UNCLOS) was produced in 1982, states taking part in the negotiations were finally able to agree that the territorial sea should be limited to 12 miles from shore.[111]

103 Ibid, p. 659.

104 Supra, Convention on the Continental Shelf, note 1.

105 'Convention on the Continental Shelf,' *United Nations Treaty Collection*, https://treaties.un.org/Pages/ViewDetails.aspx?src=TREATY&mtdsg_no=XXI-4&chapter=21&lang=en (accessed 7 February 2016). The convention achieved 43 signatories and 58 parties.

106 Tullio Treves, '1958 Geneva Conventions on the Law of the Sea,' U.N. *Audiovisual Library of International Law*, http://legal.un.org/avl/ha/gclos/gclos.html (accessed 16 November 2015).

107 Ibid.

108 Supra, Convention on the Continental Shelf, note 1, Art. 1.

109 Convention on the Territorial Sea and the Contiguous Zone, 29 April 1958, 15 U.S.T. 1606, 516 U.N.T.S. 205.

110 Supra, Treves, note 106.

111 Supra, UNCLOS, note 2, Art. 3.

The International Court of Justice

In the wake of the Truman Proclamation and with improving technology, exploration for offshore oil and gas reserves began in earnest.[112] With the uncertainties about the law of the sea that still lingered after UNCLOS I, it is not surprising that in 1967, only three years after the Convention on the Continental Shelf entered into force, the Federal Republic of Germany, Denmark and the Netherlands asked the International Court of Justice (ICJ) to render an opinion on the extent of their continental shelves after negotiations among the parties proved fruitless.[113] At the heart of the dispute was the weight to be given to Article 6 of the Convention on the Continental Shelf, which stated that, unless the parties otherwise agreed and in the absence of special circumstances, the boundary on the continental shelf between two adjacent states will be determined by drawing a line that is equidistant between them. Germany argued that application of the equidistant method would produce an unfair and inequitable result and that the equidistant method did not have the status of customary law.[114] Further, while all three states had signed the Convention on the Continental Shelf, Germany had not ratified it,[115] and thus argued that the convention's provisions were not binding on it.[116] The reason for Germany's reluctance to have the convention apply to its offshore interests originates from the negative effect that the equidistant method of Article 6(2) of the convention would have on states such as Germany that have concave coastlines.[117]

The ICJ began its legal analysis of the case with a review of the accepted fundamental principles pertaining to sovereignty over the continental shelf. Most fundamental, according to the court, is that a state's inherent sovereign right to explore the seabed and exploit the natural resources of its continental shelf derive from the state's sovereignty over the land itself, since the continental

112 *See generally*, North Sea Continental Shelf Cases (Federal Republic of Germany/Denmark; Federal Republic of Germany/Netherlands), Separate Opinion of Justice Jessup, I.C.J. Reports, 1969 ('Jessup Opinion'), *available at* http://www.icj-cij.org/docket/files/51/5543.pdf (accessed 22 November 2015).

113 North Sea Continental Shelf Cases (Federal Republic of Germany/Denmark; Federal Republic of Germany/Netherlands), Judgment, I.C.J. Reports, 1969 ('North Sea Cases'), pp. 5–6; p. 18, para. 9; p. 36, *available at* http://www.icj-cij.org/docket/files/52/5563.pdf (accessed 20 November 2015).

114 Ibid, p. 9.

115 Ibid, p. 25, para. 26.

116 Ibid, p. 20, para. 15.

117 Ibid, pp. 13–14, paras. 3–5.

shelf is a natural prolongation of the state's land territory.[118] A state's right over its continental shelf and the resources contained therein is exclusive, and no other state may explore or exploit the area without the other state's consent.[119] The court then agreed with Germany that the Convention on the Continental Shelf was not binding on that state because it had not ratified the convention, and its conduct did not support an argument of estoppel.[120]

Turning to the question of whether the equidistant method had become a tenet of customary law such that Germany was bound to utilize it,[121] the court recognized that 'no other method has the same combination of practical convenience and certainty of application.'[122] However, the court gave credence to the Truman Proclamation in acknowledging that the proclamation was widely seen as the progenitor of 'positive law' with respect to continental shelves, and its two main concepts of delimitation of boundaries through mutual agreement and equitable principles had been closely followed.[123] The UNILC had not proposed the equidistant method to delimit the boundary on the continental shelf of adjacent states until after a committee of 'hydrographical' experts suggested the concept in a report to the UNILC.[124] The ICJ suggested that the experts had proposed the equidistant method due to its practical convenience and not because of any legal grounding.[125] Therefore, the UNILC's adoption of the method for the Convention on the Continental Shelf did not represent proof of emerging customary international law[126] nor did the subsequent practice of states confer that status on the equidistant method.[127] Interestingly, not all of the convention's methods of delimitation were dismissed, for the court supported using the convention's median line method for delimitation between two states with opposite coasts.[128]

Having rejected use of the equidistant method of delimitation, the ICJ then listed the main principles that should guide states in determining boundaries on continental shelves: an obligation to negotiate with an intent to reach an agreement, an obligation to apply equitable principles, and an

118 Ibid, p. 22, para. 19; p. 29, para. 39; p. 31, para. 43.
119 Ibid.
120 Ibid, pp. 25–26, paras. 27–30.
121 Ibid, pp. 28–29, paras. 37–38.
122 Ibid, p. 23, para. 23.
123 Ibid, pp. 32–33, para. 47.
124 Ibid, p. 34, para. 50.
125 Ibid, p. 35, para. 53.
126 Ibid, p. 38, para. 62; p. 41, para. 69.
127 Ibid, p. 45, para. 81.
128 Ibid, p. 36, paras. 57–58.

acknowledgement that a state's claim to a continental shelf must only be based on the natural prolongation of its own territory, without infringement on any other state's natural prolongation.[129] With respect to the factors to be considered in delimitation, the court noted the practice of protecting the unity of natural resource deposits as a means of enabling more efficient exploitation, a practice that had already begun in the North Sea through agreements entered into during the 1960s by the United Kingdom, Norway, Germany and the Netherlands.[130] Other factors to be weighed included the configuration of coastline (taking into account any special or unusual features), the physical and geological structure of the shelf and 'a reasonable degree of proportionality.'[131]

As Jessup explained in his separate opinion, the North Sea cases were clearly about ownership of oil and gas deposits in the continental shelves.[132] In support of that observation, he made special mention of the contemporary North Sea agreements for joint exploitation of deposits[133] as well as agreements concerning deposits in the Adriatic Sea and the Persian Gulf.[134] In finding additional precedent for cooperation with respect to exploitation of natural resources, Jessup took note of the long history of states sharing water resources.[135]

Thus, in the early days of exploration and exploitation of natural resources contained in transboundary continental shelves, the North Sea cases provided guidance on several key points. States seeking to delimit the marine boundary between their adjacent continental shelves have an obligation to negotiate with a good faith intention to reach an agreement using equitable principles. Further, the natural prolongation of a state's land should guide claims of sovereignty and sovereign rights over its continental shelf. While no legal weight was given to the concept of protecting the unity of a deposit, both the ICJ judgment and Jessup's separate opinion seemed to look favorably on agreements to develop a site jointly. Significantly, the provision of Article 6(2) of the Convention on the Continental Shelf regarding the equidistant method of delimitation between adjacent states was found not to represent customary law, and the ICJ questioned whether Article 6(2) had any basis in law at all.

Two additional cases regarding delimitation on continental shelves were subsequently brought before the court. Both were decided in the 1980s, both

129 Ibid, p. 47, para. 85.

130 Ibid, pp. 51–2, para. 97.

131 Ibid, p. 54, para. 101.

132 Supra, Jessup Opinion, note 112, p. 72.

133 Ibid, p. 81.

134 Ibid, p. 82.

135 Ibid.

featured Libyan Arab Jamahiriya as a party, and both were initiated by a dispute over offshore hydrocarbons. However, one addressed the continental shelf of adjacent states, while the other addressed the continental shelf of states with opposite coasts. In its judgment in the 1982 Case Concerning the Continental Shelf between Tunisia and Libyan Arab Jamahiriya,[136] the ICJ considered the evolution and current state of international law regarding the continental shelf of adjacent states as evidenced by the Third United Nations Conference on the Law of the Sea that was then in progress. The court noted that Article 76(1) of the draft UNCLOS defined the continental shelf with reference to the 'natural prolongation' of a state's land territory.[137] Using natural prolongation as a framework for its analysis, the ICJ stated clearly that it was 'bound to decide the case on the basis of equitable principles,'[138] even though the draft UNCLOS had eliminated reference to equitable principles.[139] Neither Libya nor Tunisia was a party to the Convention on the Continental Shelf,[140] but it is significant that the ICJ judgment followed the precedent set by the North Sea cases in preferring the equitable principles method over the equidistant method of delimitation prescribed by the convention.

The 1985 Case Concerning the Continental Shelf (Libyan Arab Jamahiriya/ Malta)[141] involved two states with opposing coasts. Malta was a party to the Convention on the Continental Shelf, but, as discussed above, Libya was not a party to that convention. At the time of this ICJ case both states had signed the 1982 UNCLOS, but the treaty had not yet come into force.[142] Since Malta was a party to the Convention on the Continental Shelf, it initially proposed in 1965 to delimit the continental shelf between itself and Libya by drawing a

136 Continental Shelf (Tunisia/Libyan Arab Jamahiriya) Judgment, I.C.J. Reports 1982, p. 18, *available at* http://www.icj-cij.org/docket/files/63/6267.pdf (accessed 20 November 2015). Three years later the ICJ adjudicated another dispute between Tunisia and Libya and clarified delimitation in the first ruling. *See* Application for Revision and Interpretation of the Judgment of 24 February 1982 in the Case concerning the Continental Shelf (Tunisia/Libyan Arab Jamahiriya) (Tunisia v. Libyan Arab Jamahiriya), Judgment, I.C.J. Reports 1985, p. 192, *available at* http://www.icj-cij.org/docket/files/71/6527.pdf (accessed 25 November 2015).

137 Ibid, p. 32. para. 45.

138 Ibid, p. 44, para. 70.

139 Ibid, p. 49, para. 49.

140 Ibid, p. 43, para. 36.

141 Continental Shelf (Libyan Arab Jarnahiriya/Malta), Judgment, I.C.J. Reports 1985, p. 13, *available at* http://www.icj-cij.org/docket/files/68/6415.pdf (accessed 20 November 2015).

142 Ibid, p. 29, para, 26.

median line, in accordance with the provisions of the convention. Libya did not respond to that suggestion for eight years.[143] By the time the states requested the ICJ to adjudicate their dispute, Article 83(1) of UNCLOS had eliminated both the equidistant and median line methods of delimitation by providing that '[t]he delimitation of the continental shelf between states with opposite or adjacent coasts shall be effected by agreement on the basis of international law ... in order to achieve an equitable solution.'[144] While Article 83(1) does not mention equitable principles, both parties agreed that equitable principles must play a role in reaching an equitable solution.[145] The court recognized that the concept of the 200-mile Exclusive Economic Zone was now part of 'modern law,' and, while the EEZ principles had not 'absorbed' the continental shelf doctrines, the two concepts must be considered together.[146] Thus, natural prolongation must now be determined at least in part by distance from shore.[147] The court rejected Malta's argument that, since the two opposing coasts are fewer than 400 miles apart, the equidistant method of delimitation should be employed to produce an equitable result,[148] and provided a list of equitable principles to be utilized in delimiting maritime boundaries:

> ... the principle that there is to be no question of refashioning geography, or compensating for the inequalities of nature; the related principle of non-encroachment by one party on the natural prolongation of the other, which is no more than the negative expression of the positive rule that the coastal State enjoys sovereign rights over the continental shelf off its coasts to the full extent authorized by international law in the relevant circumstances; the principle of respect due to all such relevant circumstances; the principle that although all States are equal before the law and are entitled to equal treatment, "equity does not necessarily imply equality" (*I.C.J. Reports 1969*, p. 49, para. 91), nor does it seek to make equal what nature has made unequal; and the principle that there can be no question of distributive justice.[149]

143 Ibid, p. 28, para. 24.
144 Ibid, p. 30, para. 27.
145 Ibid, p. 31, para. 29.
146 Ibid, p. 33, para. 33.
147 Ibid, p. 33, para. 34.
148 Ibid, p. 37, paras. 42–43.
149 Ibid, pp. 39–40, para. 46.

In applying these principles to the case before it, the ICJ began with a median line between Libya and Malta and then adjusted it, giving Libya more territory to reflect its more expansive land and coastline.[150] However, the court cautioned that adoption of a median line as a starting point for delimitation 'should not be understood as implying that an equidistance line will be an appropriate beginning in all cases, or even in all cases of delimitation between opposite states.'[151]

UNCLOS

The difference between sovereignty over natural resources in the territorial sea and sovereign rights over the natural resources of the continental shelf is an important distinction that was preserved in UNCLOS. Pursuant to Article 2 of UNCLOS, a state that is a party to the treaty has total sovereignty over the waters, the airspace and the seabed and soil of its territorial sea.[152] In other words, the seabed in the territorial sea is regarded as part of the land of the applicable state and is governed by its domestic laws. By contrast, Article 56(1)(a) grants a state only sovereign rights 'for the purpose of exploring and exploiting, conserving and managing the natural resources, whether living or non-living, of the waters superjacent to the seabed and of the seabed and its subsoil.'[153] Similarly, Article 77(1) of UNCLOS grants states more limited sovereign rights over the continental shelf 'for the purpose of exploring it and exploiting its natural resources.'[154] In certain circumstances, the area of a continental shelf can be broader than that of an EEZ. Where the geological formation of the continental margin stretches further than 200 miles from the coast, the continental shelf (and the accompanying rights) under UNCLOS extend beyond the 200-mile EEZ, but not more than 350 miles from the baseline of the coast.[155]

Natural resources of the continental shelf are defined as 'the mineral and other non-living resources of the seabed and subsoil together with living organisms' that are sedentary.[156] Given the UNCLOS definition of the natural resources of the continental shelves, one must ask whether the water con-

150 Ibid, p. 56, para. 78.
151 Ibid, p. 56, para. 77.
152 Supra, UNCLOS, note 2, Art. 2.
153 Ibid, Art. 56(1)(a).
154 Ibid, Art. 77(1).
155 Ibid, Art. 76.
156 Ibid, Art. 77(4).

tained in offshore aquifers would be considered as a mineral. To be considered a mineral, a substance must have all of five characteristics: It must be naturally occurring, inorganic, and in solid form, have a limited range of chemical compositions and possess an ordered atomic structure.[157] Water in a liquid or gaseous state would not meet the definition of a mineral, although water in a frozen state—ice—could be considered as a mineral. For purposes of this analysis, an assumption will be made that water in offshore aquifers is not frozen, and thus would not be considered a continental shelf mineral within the meaning of Article 77(1).

Regarding the EEZ, Article 56 grants a state sovereign rights over natural resources,[158] but no definition limiting the scope of natural resources is provided for the exercise of those rights in the EEZ. Article 56(3) provides that '[t]he rights set out in this article with respect to the seabed and subsoil shall be exercised in accordance with Part VI,'[159] which addresses rights to the continental shelf. One interpretation of the apparent difference between Article 77 and Article 56 regarding the definition of natural resources would be that by not repeating the definition of natural resources provided in Article 77, Article 56 intended that sovereign rights in an EEZ would be broader than those in the extended continental shelf. Except for this apparent discrepancy, there is very little practical difference between the sovereign rights a state possesses in the seabed of the continental shelf and those it possesses in the seabed of the EEZ. Those rights are exclusive in that no other state may explore or exploit the natural resources without the coastal state's consent.[160]

With respect to delimitation of maritime boundaries, as the ICJ noted in the Libya-Malta case, Article 83(1) of UNCLOS directs states with opposite or adjacent coasts to reach an 'equitable solution' in fixing the limit of their shared continental shelf, and Article 74(1) uses identical language with respect to delimitation of an EEZ between states with opposite or adjacent coasts.[161] Pending finalization of an equitable solution, both Article 83(3) and Article 74(3) mandate that states 'shall make every effort to enter into

157 'Is Water a Mineral?—Is Ice a Mineral?,' Geology.com, http://geology.com/articles/water-mineral/ (accessed 7 February 2016); Edwin L. Clopton, 'Ice as a Mineral', *Rocks & Minerals*, Vol. 69(2), pp. 90–103 (1994), p. 90.

158 Supra, UNCLOS, note 2, Art. 56(1)(a).

159 Ibid, Art. 56(3).

160 Ibid, Art. 77(2).

161 Ibid, Art. 74(1).

provisional arrangements of a practical nature and, during this transitional period, not to jeopardize or hamper the reaching of the final agreement.'[162]

Hence, UNCLOS retained some of the original concepts from the Convention on the Continental Shelf, such as a state's exclusive rights to natural resources in its continental shelf. Measuring the length of the seabed over which a state has rights no longer depends exclusively on natural prolongation, although geological formations can extend a state's rights up to 350 miles from its shore. Importantly for the purposes of this inquiry, the certainty of delimiting a boundary according to an equidistant or median line has been eliminated in favor of achieving an equitable solution, which will depend on the facts and circumstances of a particular case and will most often be agreed upon by the parties themselves. Further, while attempting to find an equitable solution, states are obliged to negotiate in good faith to devise practical, temporary arrangements. The ICJ has provided guidance on the factors that should be considered in seeking an equitable solution, such as protection of the unity of a deposit and acknowledgement that equity does not mean equality.

With respect to transboundary offshore aquifers, several guidelines can be drawn from the laws of the sea. Foremost, any natural resources within a state's territorial sea, including the water in aquifers, belong to that state alone. In the event that transboundary offshore aquifers are discovered lying under the continental shelves in two or more territorial seas or EEZs, UNCLOS and the ICJ have both indicated a clear preference that the parties themselves seek and reach an equitable settlement. In the North Sea cases, ICJ Judge Jessup drew inspiration from transboundary land-based water resource management,[163] and perhaps future management of transboundary offshore aquifers would also provide inspirational precedent for sharing a vital resource. However, he also drew inspiration from William Onorato, an eminent scholar on the law of international petroleum deposits, which will be explored in the next chapter.[164]

162 Ibid, Arts. 74(3) and 83(3).

163 Supra, Jessup Opinion, note 112, p. 83.

164 Ibid, p. 84.

Part 3: Legal Principles Governing Offshore Oil and Gas Development

Background

As we have seen, the UNILC began developing a legal regime for offshore natural resources as one of its very earliest efforts, resulting in the 1958 Convention on the Continental Shelf.[165] A decade later in 1969, the ICJ issued the first of its opinions on offshore natural resources, in a case that was brought principally to determine rights to hydrocarbons lying under the North Sea.[166] The activities of the UNILC and the ICJ did not happen in a vacuum, and industrial interests did not wait for vague legal guidance from groups of international lawyers to begin crafting solutions of their own. Indeed, in the same year that the Convention of the Continental Shelf was produced, Saudi Arabia and Bahrain signed an agreement to share the income from exploitation of offshore hydrocarbons.[167] Two years later in 1960, Saudi Arabia and Kuwait jointly granted drilling concessions to a consortium of 60 Japanese companies called the Arabian Oil Company that struck oil in the Khafji field of the Persian Gulf.[168] These bi-lateral solutions, produced without direct involvement of global bodies such as the UN and the ICJ, became ever more important and necessary as each pronouncement from the official, global bodies encouraged just such an approach to resolution of disputes.

As demand for hydrocarbons increased and as drilling technology improved, states found ways to cooperate with one another where boundaries were disputed and/or where hydrocarbon deposits existed under areas over which more than one state could claim jurisdiction. Over the course of the past half-century, a fulsome body of treaties, contract law and customary practices has thus developed to guide and govern exploration and exploitation of transboundary offshore hydrocarbons. Should transboundary offshore aquifers be discovered and should their exploitation become technically feasible and economically viable, the legal regimes produced by offshore hydrocarbon exploitation would serve as useful analogous precedent, and, depending on the specific

165 Supra, Convention on the Continental Shelf, note 1.

166 Supra, North Sea Cases, note 113.

167 William T. Onorato, 'Apportionment of an International Common Petroleum Deposit,' *International & Comparative Law Quarterly*, Vol. 17, pp. 85–102 (1968), p. 87 ('Onorato 1968').

168 Fereidun Fesharaki, 'Joint Development Of Offshore Petroleum Resources: The Persian Gulf Experience?,' *Energy*, Vol. 6 (11), pp. 1325–1334 (1981), p. 1329.

treaty, may even be controlling. This chapter will explore the legal regime of transboundary hydrocarbon development, whose evolution has been chronicled by eminent scholars such as William Onorato, Masahiro Miyoshi, Rainer Lagoni and David Ong.

Early Law

Since the United States was one of the early developers of petroleum reserves, its domestic owners of hydrocarbon rights produced the first approach to transboundary development—the rule of capture. Under the rule of capture, a party with rights to the subsurface mineral had to possess the resource in order to claim ownership, leading multiple owners of rights to the same pool to engage in an extraction race with one another. Multiple wells would be drilled to tap and extract the hydrocarbons from the same resource pool, a practice that was inefficient, wasteful and conflictual.[169] According to Onorato in a 1968 article, the rule of capture was replaced by 'the doctrine of co-relative rights and duties between owners in a common source,' a doctrine by which each owner had an obligation to avoid wasting the resource intentionally or through negligence, to prevent spoilage and to extract only a fair share of the resource.[170] Onorato went on to note that in the domestic laws of most oil-producing states, co-relative rights and duties had evolved into a duty to co-operate under 'a unitized plan of development.'[171] Jessup, in his separate opinion in the ICJ's *North Sea* cases, cited Onorato's 1968 article in support of cooperative development of petroleum resources.[172] Onorato's 1968 article referred to international water law as an example of cooperation in development and use of a common resource that has properties similar to petroleum—a valuable resource with fluid nature that does not respect political boundaries.[173] Jessup noted the 'numerous examples of cooperative use and of sharing of fluvial resources' in rendering his opinion about joint exploitation of shared resources.[174] Likewise, Lagoni looked to the concept of reasonable and equitable use of the waters in a river drainage basin to support

169 Ibid, p. 89.

170 Ibid, pp. 91–92.

171 Ibid, p. 92.

172 Supra, Jessup Opinion, Note 112, p. 83. Jessup's opinion incorrectly attributes Onorato's 1968 article to 1958.

173 Supra, Onorato 1968, note 167, pp. 93, 96. Water in confined, non-recharging aquifers has additional similarities to petroleum by being encased in closed geological formations that require effort and skill to access.

174 Supra, Jessup Opinion, note 112, p. 82.

protecting the unity of a hydrocarbon deposit.[175] Twenty years after Lagoni's article was published, Ong argued in favor of protecting the unity of a deposit through unitization and joint development by citing the equitable apportionment principle of international water law which weighs the 'legitimate interests of states' in each particular set of circumstances to achieve an agreed-upon outcome.[176]

Onorato's and Jessup's opinions on the progress of a legal obligation to cooperate in the development of a common resource were probably influenced not only by customary practices for common water resources but also by a series of treaties signed in the 1960s by nations bordering the North Sea. In 1962, the Federal Republic of Germany and the Kingdom of the Netherlands supplemented a 1960 agreement on the Ems Estuary regarding protection of access to the sea by signing an additional treaty whereby the parties agreed that they and their concessionaires would cooperate on exploration and exploitation of natural resources, particularly oil and gas, in the estuary.[177] In 1965 and 1966, the United Kingdom, the Netherlands, Denmark and Norway resolved their differences over exploitation of common oil and gas deposits in the North Sea by executing several bi-lateral agreements requiring the states to attempt to reach agreement about how to exploit the resource effectively and how to apportion the costs and profits.[178] These early ventures into cooperative development of a valuable resource proved that a combination of practicality and economics could persuade states to look beyond their parochial concerns and maximize their returns through collaboration.

Unitization

In short order, the concept of developing a common resource as a single unit, known as 'unitization', was widely accepted into practice.[179] With unitization,

175 Rainer Lagoni, 'Oil and Gas Deposits Across National Frontiers,' *American Journal of International Law*, Vol. 73, pp. 215–243 (1979), pp. 239–240.

176 David M. Ong, 'Joint Development of Common Offshore Oil and Gas Deposits: "Mere" State Practice or Customary International Law?,' *The American Journal of International Law*, Vol. 93(4), pp. 771–804 (1999), p. 780 ('Ong 1999').

177 'Supplementary Agreement to the Treaty Between the Kingdom of the Netherlands and the Federal Republic of Germany Concerning Arrangements For Co- Operation In The Ems Estuary (Ems-Dollard Treaty),' *United Nations Treaty Series*, Vol. 509, pp. 140–157 (1964), *available at* https://treaties.un.org/doc/Publication/UNTS/Volume%20509/v509.pdf (last accessed 28 November 2015).

178 Supra, Onorato 1968, note 167, p. 88 and note 8.

179 Ana E. Bastida et al, 'Cross-Border Unitization and Joint Development Agreements: An International Law Perspective,' *Houston Journal of International Law*, Vol. 29, pp. 355–422, 370 (2006–2007).

common owners of an identified oil or gas field combine their separate interests into a single unit for purposes of operational exploration and extraction (often by a single operator) and also for purposes of dividing both costs and revenues.[180] The first agreement to develop an oil field jointly was signed in 1974 by Japan and South Korea, which had not yet agreed on their jurisdictional boundary but wanted to access the hydrocarbon field lying between them.[181] In 1976, Norway and the United Kingdom signed a formal international unitization agreement to develop the Frigg gas field in the North Sea, stating clearly that the Frigg Field Reservoir 'shall be exploited as a single unit.'[182] The Frigg Treaty set the standard for subsequent unitization treaties by containing provisions that were to become commonplace. Each state agreed to cause its license concessionaires to reach an agreement with the license concessionaires of the other state that would implement the provisions of the Frigg Treaty, and the resulting agreement between the concessionaires required approval of the states.[183] The states would jointly determine the limits of the field and would attempt to agree on apportionment of the reserves prior to production.[184] The license concessionaires would appoint a single unit operator,[185] landing facilities were to be freely accessible to each party,[186] the states would cooperate on safety,[187] each state retained its own taxation regime,[188] license rights could not be transferred nor could new rights be granted without approval of the other state,[189] and a special commission was established to address issues

180 James E. Horigan, 'Unitization of Petroleum Reservoirs Extending Across Sub-Sea Boundary Lines of Bordering States in the North Sea,' *Natural Resources Law*, Vol. 7, pp. 67–76, 73 (1974).

181 Masahiro Miyoshi, 'The Joint Development of Offshore Oil and Gas in Relation to Maritime Boundary Delimitation,' *International Boundaries Research Unit Maritime Briefing*, Vol. 2(5), pp. 1–51, 1 (1999) ('Miyoshi 1999').

182 'Agreement between the Government of the United Kingdom of Great Britain and Northern Ireland and the Government of the Kingdom of Norway relating to the Exploitation of the Frigg Field Reservoir and the Transmission of Gas therefrom to the United Kingdom,' Cmnd 6491, Art. 1(1) (1976) ('Frigg Treaty'), *available at* https://treaties.un.org/doc/Publication/UNTS/Volume%201098/volume-1098-I-16878-English.pdf (last accessed 28 November 2015).

183 Ibid, Art. 1(2).

184 Ibid, Art. 2.

185 Ibid, Art. 5.

186 Ibid, Art. 6.

187 Ibid, Arts. 7–8.

188 Ibid, Art. 9.

189 Ibid, Art. 10.

referred to it by the states.[190] Transmission of the gas to shore via pipelines was addressed,[191] and separate annexes listed the equipment to be installed[192] and the exact location of the four wells.[193] Subsequent agreements between the U.K. and Norway for the Statfjord, Murchison and Markham oil fields all followed the pattern of the Frigg Treaty.[194]

Customary Law

By the time that Onorato published his next paper on the topic of offshore oil and gas development in 1977, state practice with respect to commonly held offshore oil and gas resources had become so consistent that he felt confident in saying that 'there are now quite definite principles and rules of law to be applied....'[195] First, no state may unilaterally exploit a common resource over the reasonable objection of another state that has claim to that resource. Secondly, states must agree on the method and legal basis used to exploit the resource. Third, states were obliged to enter into negotiations regarding apportionment of the resource. Fourth, while international law regarding the substance of any agreement had not yet been formed, states wishing to form an agreement could look to certain other bodies of law and practice for analogies. Those other bodies of law and practice could include domestic laws requiring unitization and international laws that pertain to resources similar to petroleum or that had become *opinio juris* due to state practice with respect to petroleum.[196] Several of these principles have obvious similarities with, if not origins in, laws of the sea, such as the sovereign right of a state to exclusive development of its continental shelf natural resources and the requirement by UNCLOS and the ICJ North Sea cases to enter into negotiations with other concerned states to resolve disputed rights.

---

190 Ibid, Art. 27.

191 Ibid, Arts. 13–21.

192 Ibid, Annex A.

193 Ibid, Annex B.

194 Peter D. Cameron, 'The Rules Of Engagement: Developing Cross-Border Petroleum Deposits In The North Sea And The Caribbean,' *International & Comparative Law Quarterly*, Vol. 55, pp. 559–585, 572 (2006).

195 William T. Onorato, 'Apportionment of an International Common Petroleum Deposit,' *International & Comparative Law Quarterly*, Vol. 26, pp. 324–337, 327 (1977).

196 Ibid, pp. 327–328. Onorato restated these principles in a later article. *See* William T. Onorato, 'Joint Development of Seabed Hydrocarbon Resources: An Overview of Precedents in The North Sea,' *Energy*, Vol. 6 (11), pp. 1311–1324, 1311–1312 (1981) ('Onorato 1981').

Another legal expert on offshore oil and gas exploration, Rainer Lagoni, published an article in 1979 noting that, since the U.K.-Norway treaty was signed in 1965, many states had included a minerals deposit clause in their continental shelf delimitation treaties, obligating them to cooperate in the development of any liquid mineral deposits that are found to be commonly held resources.[197] Calling the practice 'striking in its uniformity,'[198] he suggested that the wide acceptance of mineral deposit clauses evidenced an emerging principle of customary law requiring cooperation in the development of commonly held offshore liquid mineral deposits,[199] and pointed to judicial decisions in opining that such cooperation had achieved the standard of *opinio juris* that is necessary for a practice to be considered as customary law.[200] Lagoni describes four types of cooperation agreements then in use: geological cooperation in mining the deposit;[201] joint operations by the states' concessionaires;[202] unitized development of a defined field or a common resource (unitization);[203] and 'a functionally limited condominium' whereby the states exercise joint power over the resource.[204] Like others before and after him, Lagoni referenced the international law of river basins as an analogous precedent for equitable apportionment of a common resource.[205] A few years later, Onorato declared that joint development is both economically necessary and legally required whether the marine boundary between two or more states has been clearly delimited or is in dispute.[206]

While the ICJ and UNCLOS provided limited guidance on delimiting marine boundaries that steered states to self-constructed resolutions, no such guidance

---

197 Supra, Lagoni, note 175, p. 229. As discussed earlier, water can only be viewed as a mineral when it is frozen into a solid form.

198 Ibid. In his article, Lagoni lists contemporary treaties with mineral deposit clauses covering the North Sea, Baltic Sea, the Gulf of Biscay, the Mediterranean Sea and the Adriatic Sea. Ibid, pp. 229–230.

199 Ibid, p. 233.

200 Ibid, p. 235.

201 Ibid, pp. 222–223.

202 Ibid, p. 223.

203 Ibid, p. 224.

204 Ibid, pp. 226–228.

205 Ibid, p. 236. *See also* Masahiro Miyoshi, 'Some Comments on Legal Aspects of Precedents for Joint Development,' *Energy*, Vol. 6 (11), pp. 1359–1362, 1360 (1981) ('Miyoshi 1981').

206 William T. Onorato, 'A Case Study in Joint Development: The Saudi Arabia-Kuwait Partitioned Neutral Zone,' *Energy*, Vol. 10 (3/4), pp. 539–544, 539–340 (1985).

was available for transboundary marine natural resources,[207] and states were relegated to crafting their own solutions.[208] UNCLOS clearly stated that a state retains sovereignty over the resources of its territorial sea and sovereign rights over the natural resources of the seabed of its EEZ and continental shelf.[209] By agreeing to develop a common oil or gas field in cooperation with a neighboring state, especially where that development involves establishing an oversight committee, one could argue that the state was conceding some of its sovereign rights to another state or to the committee. This issue was noted even as early as the Frigg Treaty, and to protect themselves against such claims, states entering into a unitization arrangement to develop an identified common field specifically preserved their rights under international law to resources in the continental shelf.[210] Given the vast sums of money that accrued from offshore oil and gas development, taking the small risk that treaty language would not completely protect sovereignty over natural resources must have seemed like a wise wager to make.[211]

Joint Development Agreements

As described above, unitization of common oil and gas fields became commonplace in a relatively short period of time, but the practice addressed only exploitation after a field had been identified and designated. The next legal structure to embody a communal approach to development evolved from the desire to cooperate on exploration as well as exploitation. The structure of a joint development agreement (JDA) allowed the parties to share costs and risk in the early stages of developing a field while searching for oil and gas reserves prior to exploitation. Generally speaking, a unitization agreement formalizes joint exploitation of a known oil or gas pool, whereas in a JDA the state parties select an area where oil and/or gas reserves are expected to be found and agree to joint exploration and exploitation.[212] The maritime boundaries between the states need not be finally agreed-upon in order for them to

207 Alberto Szekely, 'The International Law of Submarine Transboundary Hydrocarbon Resources: Legal Limits to Behavior and Experiences for the Gulf of Mexico,' *Natural Resources Journal*, Vol. 26, pp. 733–768, 738, 743 (1986).

208 Ibid, p. 758.

209 Supra, UNCLOS, note 2.

210 Supra, Frigg Treaty, note 182, Art. 29(1). *See also* supra, Onorato 1981, note 196, p. 1314.

211 Supra, Miyoshi 1981, note 205, p. 1359.

212 Supra, Ana E. Bastida et al, note 179, pp. 358–359.

cooperate in developing common offshore resources through a JDA.[213] In 1999, Ong described three types of JDAs: (1) one state performs the development and the other state shares in the net profits after costs are deducted; (2) the states form a sort of joint venture, with each of them involved in approving any joint operating agreements and with the joint commission having limited powers, and (3) the states delegate much of the development responsibility and operation to a strong joint commission.[214] Since use of JDAs was quickly and widely adopted in certain parts of the world, Ong argued that a regional rule of customary international law had arisen requiring JDAs to be formed in those areas where they were most regularly found, such as the North Sea, the Persian Gulf and Southeast Asia.[215] While the obligation to create a JDA may be only a regional requirement, there remained firm rules of customary law applicable to all shared resources for states to cooperate in trying to craft an agreement and to share information.[216] Exploration could proceed unilaterally, assuming that the activity caused no harm to the other state's sovereign interests, but exploitation of a natural seabed resource had to be agreed upon by the relevant states,[217] thus putting the rule of capture firmly to rest.[218]

In a 2002 article, Ong cited Onorato and M.J. Valencia in listing the components of a traditional JDA:

> (i) a treaty designating the geographical area subject to the joint development regime, (ii) the establishment of some form of international joint body, composed of equal representatives from both parties, (iii) the grant of exploration and exploitation rights over the designated joint development area, either by the parties or the established joint body, and (iv) the nomination by the individual rights holders of a single operator to exploit the deposits in the designated area.[219]

213 Paul Michael Blyschak, 'Offshore Oil and Gas Projects amid Maritime Border Disputes: Applicable Law,' *Journal of World Energy Law and Business*, Vol. 6(3), pp. 210–233, 217 (2013).

214 Supra, Ong 1999, note 176, pp. 788–791.

215 Ibid, p. 795.

216 Ibid, p. 798.

217 Ibid, p. 800.

218 Ibid, p. 802.

219 David M. Ong, 'The New Timor Sea Arrangement 2001: Is Joint Development of Common Offshore Oil and Gas Deposits Mandated under International Law?,' *International Journal of Marine & Coastal Law*, Vol. 17, pp. 79–122, 95 (2002).

Given the positive practical and legal characteristics of JDAs, at least a dozen were completed and implemented around the world, both where maritime boundaries were delimited and where they were still in dispute. In 1999, Miyoshi compiled a list of JDAs then in force:[220]

JDAs where the maritime boundary was not delimited

- 1974 Japan–South Korea Joint Development Agreement[221]
- 1979 Malaysia–Thailand Memorandum of Understanding[222]
- 1989 Timor Gap Treaty between Australia and Indonesia[223]
- 1992 Malaysia–Vietnam Memorandum of Understanding[224]
- 1993 Colombia–Jamaica Treaty[225]
- 1995 Argentina–United Kingdom Joint Declaration[226]

---

220 Supra, Miyoshi 1999, note 181, pp. 6–39.

221 Agreement between Japan and the Republic of Korea concerning Joint Development of the Southern Part of the Continental Shelf Adjacent to the Two Countries, 30 January 1974, 1225 UNTS 114, *available at* https://treaties.un.org/doc/Publication/UNTS/Volume%20 1225/volume-1225-I-19778-English.pdf (accessed 3 December 2015).

222 1979 Memorandum of Understanding between Malaysia and the Kingdom of Thailand on the Establishment of the Joint Authority for the Exploitation of the Resources of the Seabed in a Defined Area of the Continental Shelf of the Two Countries in the Gulf of Thailand, 21 February 1979, *available at* http://cil.nus.edu.sg/rp/il/pdf/1979%20MOU%20 between%20Malaysia%20and%20Thailand-pdf.pdf (accessed 3 December 2015).

223 Agreement between the Government of Australia and the Government of the Democratic Republic of Timor-Leste relating to the Unitization of the Sunrise and Troubadour Fields, 6 March 2003, 2483 UNTS 317, *available at* http://www.un.org/Depts/los/ LEGISLATIONANDTREATIES/PDFFILES/TREATIES/AUS-TLS2003UNI.PDF (accessed 3 December 2015).

224 1992 Memorandum of Understanding between Malaysia and the Socialist Republic of Vietnam for the Exploration and Exploitation of Petroleum in a Defined Area of the Continental Shelf Involving the Two Countries, 5 June 1992, *available at* https://assets .documentcloud.org/documents/1352723/vtn-mal-mou-between-malaysia-and-vietnam-for-the.pdf (accessed 7 February 2016).

225 Maritime delimitation treaty between Jamaica and the Republic of Colombia, 12 November 1993, Reg. No. I-30943, *available at* http://www.un.org/Depts/los/ LEGISLATIONANDTREATIES/PDFFILES/TREATIES/JAM-COL1993MD.PDF (accessed 7 February 2016).

226 Joint Declaration of 27 September 1995 on Cooperation over Offshore Activities in the South West Atlantic, *available at* http://fiassociation.com/shopimages/pdfs/7.%201995%20 Joint%20Declaration%20on%20Cooperation%20Over%20Offshore%20Activities%20 in%20the%20South%20West%20Atlantic..pdf (accessed 3 December 2015). In 2007, Argentina withdrew from the cooperation agreement. *See* Duncan Campbell, 'Falklands

JDAs where the maritime boundary was delimited

- 1974 France–Spain Agreement[227]
- 1974 Saudi Arabia–Sudan Agreement[228]
- 1981 Iceland–Norway Agreement[229]
- 1988 Libya–Tunisia Agreement[230]
- 1993 Guinea-Bissau–Senegal Agreement[231]

Since publication of Miyoshi's analysis, the 2001 Nigeria–São Tomé and Príncipe Joint Development Agreement has also been completed,[232] as has the 2012 Agreement between the United States of America and the United Mexican States Concerning Transboundary Hydrocarbon Reservoirs in the Gulf of Mexico.[233] Some of these JDAs have been agreed upon in regions

gulf Widens as Argentina Quits Oil Deal with UK,' *The Guardian* (29 March 2007), http://www.theguardian.com/uk/2007/mar/29/argentina.falklands (accessed 3 December 2015).

227 Convention between the Government of the French Republic and the Government of the Spanish State on the Delimitation of the Continental Shelves of the two States in the Bay of Biscay (Golfe de Gascogne/Golfo de Vizcaya), 29 January 1974, *available at* http://www.un.org/Depts/los/LEGISLATIONANDTREATIES/PDFFILES/TREATIES/FRA-ESP1974CS.PDF (accessed 3 December 2015).

228 Agreement between Sudan and Saudi Arabia Relating to the Joint Exploitation of the Natural Resources of the Seabed and Subsoil of the Red Sea in the Common Zone, 16 May 1974, 952 UNTS 197.

229 Agreement on the Continental Shelf Between Iceland and Jan Mayen, 22 October 1981, *available at* http://www.un.org/Depts/los/LEGISLATIONANDTREATIES/PDFFILES/TREATIES/ISL-NOR1981CS.PDF (last accessed 3 December 2015).

230 Agreement between the Libyan Arab Socialist People's Jamahariya and the Republic of Tunisia to Implement the Judgment of the International Court of Justice in the Tunisia/Libya Continental Shelf Case, 8 August 1988, *available at* http://www.un.org/Depts/los/LEGISLATIONANDTREATIES/PDFFILES/TREATIES/LBY-TUN1988CS.PDF (last accessed 3 December 2015).

231 Management and Cooperation Agreement between the Government of the Republic of Senegal and the Government of the Republic of Guinea-Bissau, 14 October 1993, *available at* http://www.un.org/Depts/los/LEGISLATIONANDTREATIES/PDFFILES/TREATIES/SEN-GNB1993MC.PDF (last accessed 3 December 2015).

232 Treaty between the Federal Republic of Nigeria and the Democratic Republic of São Tomé and Príncipe on the Joint Development of Petroleum and other Resources, in respect of Areas of the Exclusive Economic Zone of the Two States, 21 February 2001, *available at* http://www.un.org/Depts/los/LEGISLATIONANDTREATIES/PDFFILES/TREATIES/STP-NGA2001.PDF (last accessed 3 December 2015).

233 Agreement between the United States of America and the United Mexican States Concerning Transboundary Hydrocarbon Reservoirs in the Gulf of Mexico, Feb. 20, 2012, T.I.A.S. No. 14–718, *available at* http://www.state.gov/p/wha/rls/2012/185259.htm.

outside of the areas identified by Ong in 1999—the North Sea, the Persian Gulf and Southeast Asia—which raises a question as to whether the regional rule of customary law now has broader application.

Framework Agreement

As described above, unitization agreements providing for exploitation of a single hydrocarbon deposit evolved into JDAs whose provisions allowed for exploration and exploitation of reserves in a designated area and that were flexible enough not to require that the parties agree on the boundary of their shared continental shelf. The next step in the evolution of treaties for development of offshore resources was once again introduced by the United Kingdom and Norway for North Sea hydrocarbon deposits, through the mechanism of the 2005 Framework Agreement concerning Cross-Boundary Petroleum Co-operation ('Framework Agreement').[234] With this agreement, the U.K. and Norway departed from the JDA pattern whereby each field was the subject of a separate agreement. Instead, the two states agreed on terms that would apply to any designated 'Cross-Boundary Project,' which could include construction and operation of a cross-boundary pipeline, exploration for and exploitation of a transboundary oil or gas field, construction and operation of an installation for use in exploration and/or exploitation, and use of what are termed 'Host Facilities' that are utilized by the parties to explore and develop fields outside of the geographical limits of the agreement.[235]

In many respects, the provisions of the Framework Agreement bear resemblance to JDAs, with states agreeing to use 'best efforts to facilitate Cross-Boundary Projects,'[236] to exchange information,[237] to form a committee to facilitate the agreement,[238] to grant necessary authorizations,[239] to require their respective licensees to reach an agreement regulating the trans-boundary reservoir[240] and to appoint a single operator.[241] The licensee agreement must be submitted to both governments, and, to streamline the process, the license

234 Framework Agreement between the Government of the United Kingdom of Great Britain and Northern Ireland and the Government of the Kingdom of Norway Concerning Cross-Boundary Petroleum Co-operation, 4 April 2005, *available at* https://www.gov.uk/government/uploads/system/uploads/attachment_data/file/273282/6792.pdf (last accessed 3 December 2015).

235 Ibid, Art. 1.2.

236 Ibid, Art. 1.4(1).

237 Ibid, Art. 1.10.

238 Ibid, Art. 1.15.

239 Ibid, Arts. 2.1(1) and 3.1(2).

240 Ibid, Art. 3.2(1).

241 Ibid, Art. 3.7.

agreement is deemed to be approved unless the licensees are notified otherwise within 60 days of receipt by the government of the agreement.[242] The appointment of the initial unit operator and any replacement requires approval of both governments.[243] Once again, to avoid any doubt on the matter, the states declare that the Framework Agreement will not affect their sovereign rights to the continental shelf under international law.[244] Offshore pipelines and hydrocarbon development projects covered under six prior treaties ranging from 1973 to 1998, including the iconic Frigg Treaty, are specifically excluded from the Framework Agreement.[245] Two small oil fields and a gas transmission project were coupled with the Framework Agreement at the time of its announcement.[246] Since the Framework Agreement is so recent, not enough time has passed to know whether its comprehensive and practical approach to the development of offshore resources will become an accepted trend. The framework structure may indeed represent a new paradigm of pre-negotiating terms and conditions and then assigning identified fields, or it may eventually simply reflect market terms in 2005.

In summary, the customary laws and practices of offshore hydrocarbon development have evolved and solidified with guidance from ICJ and UNCLOS pronouncements. Even in the absence of fully delimited boundaries, states enter into agreements to collaborate on exploration and exploitation of commonly held fields. These agreements generally designate a geographical area to fall within the scope of the agreement and provide that the states will cause a single operator to develop a designated field or fields. This approach has obvious practical benefits, since use of a single operator eliminates any risk of a drilling race to the bottom of the resource pool by competing developers. In addition, sharing of data and information through the mechanism of a single operator aids in maximizing efficient extraction. Concerns about waivers of sovereign rights are addressed in the agreements by preserving all rights under international law. These practices have arisen in an attempt to streamline the development of economically valuable resources that are located in challenging offshore environments in geological formations that are susceptible to contamination and inefficient depletion from uncoordinated activities.

The philosophy of these offshore oil and gas development practices drew initial inspiration from joint development of land-based water resources and

242 Ibid, Art. 3.2(2).
243 Ibid, Art. 3.7.
244 Ibid, Art. 1.3.
245 Ibid, Art. 1.2. and Annex E.
246 Supra, Cameron, note 194, p. 576.

then evolved in their own unique manner. In order to better understand these origins and suggest a legal regime for transboundary offshore aquifers, the next chapter will describe the history and current state of the laws and practices of both transboundary land-based surface water and transboundary land-based aquifers.

Part 4: Legal Principles Governing Land-based Aquifers

Foundations of Aquifer Governance

Surface water in lakes and rivers has been the topic of discussions and agreements among nations for centuries, and thus an extensive and textured body of treaty and customary law has developed to guide the equitable use of those water resources. Indeed, as discussed earlier, the legal principles governing transboundary surface water have been so widely accepted that scholars attempting to design a legal regime for commonly held offshore hydrocarbon reserves looked to these well-established principles for guidance. Both nations and courts recognize the basic principles that have come to govern utilization of transboundary surface water resources: equitable and reasonable use, no significant harm to the resource, cooperation, sharing of information, protection of the environment and dispute resolution.[247] Owen McIntyre suggests that these and other principles governing water use can be classified as either 'substantive' or 'procedural':

> These substantive rules include the duty to prevent transboundary pollution, the duty to co-operate, the duty to conduct transboundary environmental impact assessments, the doctrine of sustainable development, the principle of intergenerational equity, the principle of common but differentiated responsibility, the precautionary principle, the polluter pays principle, and the ecosystems approach. The procedural rules include the duty to notify, duties relating to the ongoing exchange of information, the duties to consult and to negotiate in good faith, the duty to warn, and duties relating to the settlement of disputes.[248]

---

247 L. Boisson de Chazournes, *Fresh Water in International Law* (Oxford University Press 2013), pp. 30–32.

248 Owen McIntyre, 'The Role of Customary Rules and Principles of International Environmental Law in the Protection of Shared International Freshwater Resources,' *Natural Resources Journal*, Vol. 46, pp. 157–210 (2006), p. 159.

Like rivers and lakes, groundwater does not respect state boundaries, and flows where the rock formations and pressure differentials may lead it. Perhaps due to the inscrutable subterranean nature of transboundary aquifers and the difficulties of monitoring withdrawals and capacity, management of shared groundwater has not benefitted from much legal attention until recently, even though the volume of groundwater on the planet far surpasses the volume of surface water. According to a 1993 report, only 2.5% of the world's water is fresh water, and of that small percentage, 68.7% is frozen in polar caps and glaciers and is thus unavailable for human use. Groundwater holds another 30.1% of the world's fresh water, with surface water accounting for only 1.2%.[249] Since the date of that report, vast quantities of groundwater have been drawn to the surface—an estimated one-third to 43% of all water used for irrigation globally originated in aquifers.[250] Nevertheless, in 2005 the Global Environment Facility estimated that groundwater still comprises 98% of the available fresh water on the planet.[251]

Not all aquifers are created equal, and the differences were ably described in a 2003 article by father and son Yoram and Gabriel Eckstein. In basic terms, aquifers exist in geological formations whose characteristics cause the aquifers to be classified as either 'unconfined' or 'confined.' In both cases, the aquifer sits atop an underlying layer of impermeable material, through which the water cannot flow any deeper. Unconfined aquifers have a layer of permeable material above them that allows water (from rivers, lakes, rain, etc.) to seep into the aquifer formation and recharge its contents. Confined aquifers are sandwiched between two layers of impermeable material. Some confined aquifers are so confined by impermeable material that they cannot be recharged in any meaningful timeframe. Others, however, can be recharged, either by another aquifer, and/or by seepage of water from the surface into the aquifer's recharge zone. One type of aquifer that receives almost no contemporary recharge is known as a fossil aquifer. These aquifers, which are usually found in arid

---

249 Igor A. Shiklomanov, 'World Fresh Water Resources,' *in* Peter . Gleick, Editor, *Water in Crisis: A Guide to the World's Fresh Water Resources*, pp. 13–24 (Oxford University Press 1993), p. 13.

250 Tushaar Shah, 'Groundwater Governance and Irrigated Agriculture,' Technical Paper No. 19, *Global Water Partnership* (2014) p. 8, *available at* http://www.gwp.org/Global/ToolBox/Publications/Background%20papers/GWP_TEC_19_web.pdf (accessed 5 December 2015) (using the figure of one-third). *What is Groundwater?*, International Groundwater Resources Assessment Center, http://www.un-igrac.org/what-groundwater (accessed 5 December 2015) (using figure of 43%).

251 'Medium Sized Project Proposal,' *Global Environment Facility* (2005) ('GEF 2005 Proposal'), p. 6, *available at* http://www-naweb.iaea.org/napc/ih/documents/Nubian/Nubian_final_MSP_Sandstone.pdf (accessed 6 December 2015).

environments and receive little or no recharge, typically contain water that is thousands to millions of years old.[252] As described in Chapter 1, although some offshore aquifers are hydraulically connected to, and thus recharged by, a land-based aquifer, no offshore aquifer receives fresh water recharge through an overlying permeable geological formation. Therefore, all offshore aquifers can be classified as confined aquifers, and those containing palaeowater can be considered as fossil aquifers.

Surprisingly, of the nearly 600 identified transboundary aquifers,[253] only four benefit from fully ratified agreements: the Franco-Genevese Aquifer, which is common to France and Switzerland;[254] the Nubian Sandstone Aquifer System, which is common to Libya, Sudan, Egypt and Chad;[255] the Northwest Sahara Aquifer System which is common to Algeria, Libya and Tunisia;[256] and most recently in 2015, the Al-Sag/Al-Disi Aquifer, which is common to Saudi Arabia and Jordan.[257] In addition, in 1973 the U.S. and Mexico agreed in Minute

---

252 Gabriel Eckstein & Yoram Eckstein, 'A Hydrogeological Approach to Transboundary Ground Water Resources and International Law,' *American University International Law Review*, Vol. 19, pp. 201–258 (2003), pp. 210–217.

253 'Special Edition of TBA Map will be launched at WWF7' (April 7, 2015), International Groundwater Resources Assessment Center, http://www.un-igrac.org/news/special-edition-tba-map-will-be-launched-wwf7 (accessed 5 December 2015).

254 Convention relative à la protection, à l'utilisation, à la réalimentation et au suivi de la nappe souterraine franco-suisse du Genevois ("2007 Franco-Genevese Aquifer Convention"), EIF 1 January 2008, the Community of the 'Annemassienne' Region, the Community of the 'Genevois' Rural Districts, and the Rural District of Viry and the Republic and Canton of Geneva, *available at* http://internationalwaterlaw.org/documents/europe.html#Franco-Swiss%20Genevese%20Aquifer (accessed 6 December 2015).

255 Supra, GEF 2005 Proposal, note 251, Annex 7, 'Constitution of the Joint Authority for the Study and Development of the Nubian Sandstone Aquifer Waters, 1992, Libya, Chad, Sudan and Egypt ('NSAS 1992 Constitution'); Programme for the Development of a Regional Strategy for the Utilisation of the Nubian Sandstone Aquifer System (NSAS)—Terms of Reference for the Monitoring and Exchange of Groundwater Information of the Nubian Sandstone Aquifer System and Terms of Reference for Monitoring and Data Sharing ('NSAS 2000 Agreements'), 5 October 2000, Libya, Chad, Sudan and Egypt, *available at* http://www.fao.org/docrep/008/y5739e/y5739e05.htm (accessed 5 December 2015).

256 Establishment of a Consultation Mechanism for the Northwestern Sahara Aquifer System (SASS), 2002, Algeria, Libya and Tunisia ('NWSAS Agreement'), *available at* http://www.fao.org/docrep/008/y5739e/y5739e05.htm (accessed 27 December 2015).

257 Agreement between the Government of the Hashemite Kingdom of Jordan and the Government of the Kingdom of Saudi Arabia for the Management and Utilization of the Ground Waters in the Al-Sag/Al-Disi Layer, 30 April 2015, Saudi Arabia and Jordan ('Al Sag/Al-Disi Agreement'), *available at* http://internationalwaterlaw.org/documents/asia.html#Al-Sag/Al-Disi%20Aquifer (accessed 5 December 2015).

242 of the International Boundary & Water Commission to limit groundwater pumping along a small section of the U.S.-Mexico border near San Luis, Arizona and to consult with one another regarding future development that would affect the surface or groundwater.[258] Several other aquifers have been the subject of negotiated draft agreements that are not yet fully in force—the Iullemeden Aquifer, which is common to Mali, Niger and Nigeria;[259] the Guarani Aquifer System, which is common to Brazil, Argentina, Uruguay and Paraguay;[260] and the Iullemeden, Taoudeni/Tanezrouft Aquifer System, which is common to Algeria, Benin, Burkina Faso, Mali, Mauritania, Niger and Nigeria.[261] Nevertheless, the provisions of both sets of agreements can be analyzed as evidence of emerging customary law on transboundary aquifers.

Aquifer Agreements

The initial Franco-Genevese Aquifer Agreement[262] entered into force in 1977 and presented a very practical method for two sub-national political units—

---

258 Permanent and Definitive Solution to International Problem of the Salinity of the Colorado River, Minute 242 of the International Boundary & Water Commission, para. 5, 6, 30 August 1973, *available* at http://www.ibwc.gov/Files/Minutes/Min242.pdf (accessed 7 February 2016). Minutes of the International Boundary & Water Commission represent amendments to a 1944 treaty between the U.S. and Mexico. 'Treaty Relating to the Utilization of Waters of the Colorado and Tijuana Rivers and of the Rio Grande, 59. Stat 1219, T.S. 994, 9 Bevins 1166, 3 U.N.T.S. 313 (3 February 1944), *available at* http://www.ibwc.state.gov/Files/1944Treaty.pdf (accessed 7 February 2016).

259 Memorandum of Understanding Relating to the Setting Up of a Consultative Mechanism for the Management of the Iullemeden Aquifer System, 20 June 2009, Mali, Niger and Nigeria ('2009 Iullemeden MOU'), *available at* http://www.internationalwaterlaw.org/documents/regionaldocs/Iullemeden_MoU-2009.pdf (accessed 5 December 2015).

260 Agreement on the Guarani Aquifer, 2 August 2010, Brazil, Argentina, Uruguay and Paraguay ('2010 Guarani Aquifer Agreement'), *available at* http://internationalwaterlaw.org/documents/s-america.html (accessed 5 December 2015).

261 Memorandum of Understanding for the Establishment of a Consultation Mechanism for the Integrated Management of the Water Resources of the Iullemeden, Taoudeni/Tanezrouft Aquifer Systems (ITAS), 28 March 2014, Algeria, Benin, Burkina Faso, Mali, Mauritania, Niger and Nigeria ('2014 ITAS MOU'), *available at* http://internationalwaterlaw.org/documents/africa.html#Nubian%20Sandstone%20Aquifer (accessed 5 December 2015).

262 Arrangement relatif à la Protection, à l'Utilization et à la Réalimentation de la Nappe souterraine franco-suisse du Genevois, EIF 9 June 1978, the State Council of the Republic and Canton of Geneva and the Prefect of Haute-Savoie, *available at* http://www.internationalwaterlaw.org/documents/regionaldocs/franko-swiss-aquifer.html (accessed 6 December 2015).

the Swiss State Council of the Republic and Canton of Geneva and the French Prefect of Haute-Savoie—to share water resources from an aquifer lying under the southeast rim of Lake Geneva (also called Lac Léman). The parties established a management commission[263] to propose a utilization program regarding withdrawals from and protection of the aquifer,[264] to take inventory of all aquifer-related equipment,[265] to give a technical opinion on all new waterworks[266] and to monitor the quality of the water being extracted and injected.[267] Functionally, the parties agreed that the Republic and Canton of Geneva would be responsible for construction and operation of a facility that would recharge the aquifer with water.[268] The French prefect was given the right to withdraw limited quantities of water[269] in return for sharing in the costs of construction and operation of the artificial recharge installation.[270] This agreement was designed to terminate in 30 years[271] but could automatically renew for five-year periods unless one party served a one-year notice of termination.[272] In 2007, the Republic and Canton of Geneva signed another agreement with slightly different French subnational political units,[273] and the provisions of this new agreement are largely similar to those of the original agreement, including the volume of water permitted to the French partners,[274] the formula for cost-sharing[275] and the 30-year term,[276] with automatic renewals every three years instead of five years.[277] The 2007 agreement contains a new provision requiring the parties to share data.[278] Since no other multinational agreement involves artificial recharge, most of the provisions of the

263 Ibid, Art. 1.
264 Ibid, Art. 2.
265 Ibid, Art. 4.
266 Ibid, Art. 5.
267 Ibid, Art. 16.
268 Ibid, Art. 8. Geneva was permitted to 'entrust' a third party to operate the installation.
269 Ibid, Art. 9.
270 Ibid, Art. 14.
271 Ibid, Art. 19(1).
272 Ibid, Art. 19(2).
273 The 1977 agreement featured the Prefect of Haute-Savoie as the French party, and the 2007 Franco-Swiss Aquifer Convention featured the Community of the 'Annemassienne' region, the Community of the 'Genevois' Rural Districts, and the Rural District of Viry as French parties.
274 Supra, 2007 Franco-Genevese Aquifer Convention, note 254, Art. 8.
275 Ibid, Art. 14.
276 Ibid, Art. 19(1).
277 Ibid, Art. 19(2).
278 Ibid, Art. 10.

Franco-Genevese aquifer agreements have not been replicated in subsequent treaties elsewhere in the world. The principal value of the Franco-Genevese agreements stems from the fact that the 1977 agreement was the first to focus solely on an aquifer, and thus together the two agreements concerning the Franco-Genevese aquifer provide a solid and practical example of effective long-term cross-border cooperation in managing an aquifer resource.

The next aquifers to be the subject of multi-national treaties are both fossil aquifers lying under deserts in northern Africa. The Nubian Sandstone Aquifer System (NSAS) is one of the largest aquifer systems in the world and lies underneath an area of approximately two million square kilometers.[279] Discovered in 1953 during the search for oil, the NSAS's four major groundwater basins contain vast quantities of water that were emplaced thousand of years ago.[280] In 1992, Egypt and Libya established a joint authority that was given a number of responsibilities, including collection of all information and data developed by the member states, preparation of any other studies necessary to have a full understanding of the quality and quantity of the water in the NSAS, and development of programs for resource utilization and 'rational consumption' by member states.[281] Sudan joined in 1996, and Chad followed in 1999.[282] The joint authority was designed to have corporate status,[283] be led by an executive director[284] with an administrative body[285] and a board of directors,[286] and have its expenses shared equally among the parties.[287] In 2000, the four states signed two further agreements: the Terms of Reference for the Monitoring and Exchange of Groundwater Information of the Nubian Sandstone Aquifer System, and the Terms of Reference for Monitoring and Data Sharing.[288] Hence, the agreements for the NSAS primarily focus on data sharing and not on active management of the resource.

The other fossil aquifer in northern Africa that attracted enough attention to warrant a treaty is the Northwest Sahara Aquifer System (NWSAS), lying under

279 Supra, GEF 2005 Proposal, note 251, p. 3.

280 'GMR (Great Man-Made River) Water Supply Project, Libya,' water-technology.net, http://www.water-technology.net/projects/gmr (accessed 6 December 2015).

281 Supra, NSAS 1992 Constitution, note 255, Art. 3.

282 Supra, GEF 2005 Proposal, note 251, p. 4.

283 Ibid, Art. 24.

284 Ibid, Art. 13.

285 Ibid, Art. 14.

286 Ibid, Art. 5.

287 Ibid, Art. 16.

288 Supra, NSAS 2000 Agreements, note 255.

Libya, Algeria and Tunisia. In 2002, the states that are common to the NWSAS established a consultation mechanism whose stated objective was 'to coordinate, promote and facilitate the rational management of the NWSAS water resources.'[289] Administered by the Observatory of the Sahara and Sahel,[290] the consultative mechanism was instructed to develop, gather, analyze and update data on the aquifer,[291] but, similar to the arrangement for the NSAS, was given no actual management authority.

The agreement that Saudi Arabia and Jordan concluded in 2015 regarding the Al-Sag/Al-Disi Aquifer, which receives minimal recharge, goes further than the agreements for the NSAS and the NWSAS in that it establishes a protected area within which no extraction activities may take place for a period of five years.[292] In the larger management area, the parties agree to protect the aquifer from pollution,[293] to establish technical standards for drilling[294] and to utilize the extracted water only for municipal purposes.[295] Similar to the other aquifer agreements, the Al-Sag/Al-Disi agreement creates the Joint Technical Committee,[296] which is charged with collecting and submitting data and information to the states[297] and also with monitoring quantities of withdrawals.[298]

Thus, the aquifer agreements currently in force all provide for data exchange on the aquifer and establish a committee with limited powers. For two aquifers—the Franco-Genevese and the Al-Sag/Al-Disi—withdrawals are monitored, but, with the exception of the Franco-Genevese Aquifer, the treaties do not go so far as to give management or operational powers to their constituted committees.

There are several, much more robust multilateral agreements that have been negotiated and drafted but that have not yet entered into force. For example, in 2009, Mali, Niger and Nigeria negotiated a memorandum of understanding (MOU) with respect to the Illumeden aquifer system, which consists of two main aquifers—the lower Continental Intercalaire and the

289 Supra, NWSAS Agreement, note 256, (i).

290 The website for the Observatory of the Sahara and Sahel is http://www.oss-online.org/en (accessed 23 December 2015).

291 Ibid, (iv).

292 Supra, Al-Sag/Al-Disi Agreement, note 257, Art. 2(1).

293 Ibid, Art. 2(5).

294 Ibid, Art. 2(4).

295 Ibid, Art. 2(6).

296 Ibid, Art. 3(1).

297 Ibid, Art. 3(4)(c).

298 Ibid, Art. 3(4)(b).

upper Continental Terminal.[299] Like the in-force agreements discussed above, this MOU established a consultative mechanism whose scope of duties also included monitoring, coordinating and reporting,[300] but in this case the consultative mechanism was to be given legal status enabling it to enter into contracts and acquire property[301] and was to consist of a Council of Ministers, as well as an *ad hoc* Technical Committee of Experts, National Technical and Scientific Committees and an Executive Secretariat.[302] The MOU went further in also including provisions mandating 'consideration [of] the principles' of equitable and reasonable use,[303] public participation,[304] non-detrimental use,[305] precautionary measures,[306] polluter-pays[307] and user-pays.[308] In addition, the draft requires data-sharing,[309] prior notification of any activity that could adversely affect the aquifer,[310] environmental protection of ecosystems and conservation of aquifer resources.[311] As stated earlier, all of these principles can be considered as part of the body of customary law of surface waters.

In 2014, five years after negotiation of the Iullemeden MOU and reportedly at the suggestion of the development partner,[312] the original parties to that MOU joined with Algeria, Benin, Burkina-Faso and Mauritania to adopt another MOU that established a consultative mechanism whose scope of responsibility extended to both the Iullemeden and the Taoudeni/Tanezrouft Aquifer System, a large aquifer[313] that is hydraulically connected to the Iullemeden. The two aquifer systems were given the acronym 'ITAS,' and the resulting MOU resembled the 2009 Iullemeden MOU in most respects,

299 Supra, 2009 Iullemeden MOU, note 259, Art. 1.
300 Ibid, Art. 5.
301 Ibid, Art. 6.
302 Ibid, Art. 7.
303 Ibid, Art. 13.
304 Ibid, Art. 14.
305 Ibid, Art. 15.
306 Ibid, Art. 16.
307 Ibid, Art. 17.
308 Ibid, Art. 18.
309 Ibid, Art. 19(a).
310 Ibid, Arts. 20, 22–28.
311 Ibid, Art. 21.
312 Chidimma C. Okeke, 'Nigeria, Other African Countries Intensify Efforts on Groundwater Resources,' *Daily Trust* (2 April 2014), *available at* http://www.dailytrust.com.ng/daily/index.php/environment/20541-nigeria-other-african-countries-intensify-efforts-on-groundwater-resources (accessed 10 February 2016).
313 The Taoudeni/Tanezrouft Aquifer System lies under an area of approximately two million square kilometers. Supra, 2014 ITAS MOU, note 261, Art. 1.

except that new principles were added to the ones listed above: a general duty to cooperate,[314] complementarity,[315] sustainable development,[316] partnership[317] and planning.[318] While these additional principles would result in better management of the aquifer resource, with the exception of the general duty to cooperate they do not as yet represent tenets of international customary water law. As stated earlier, the ITAS MOU has not yet been fully executed or ratified, so the MOU does not add flesh to the bare bones of the customary international law of aquifers.

All of the agreements and MOUs discussed above related to aquifers in Africa or the Middle East. In 2010, a large Latin American aquifer, the Guarani, became the subject of its own draft agreement. The state parties—Brazil, Argentina, Paraguay and Uruguay—declared themselves to be the 'sole owners' of the resource[319] and agreed that '[e]ach Party exercises sovereign territorial control over their respective portions of the Guarani Aquifer System.'[320] While stating that each party has 'the sovereign right to promote the management, monitoring, and sustainable utilization of the . . . water resources,' the states agree to follow the principle of reasonable and sustainable utilization,[321] to cause no significant harm to the aquifer or the environment,[322] to share technical information,[323] to give one another prior notice of activities that may have an impact on the aquifer,[324] to settle disputes amicably[325] and to use another treaty they have in common—the Treaty of the Plata River Basin—to establish a commission to coordinate their cooperation and compliance under Guarani Aquifer agreement.[326] While at the same time protecting the sovereignty of the member states, the provisions of this draft agreement conform to basic principles of international customary law of surface water. However, as with the Iullemeden and ITAS MOUs, the Guarani Aquifer agreement has not

314 Ibid, Art. 14.
315 Ibid, Art. 15.
316 Ibid, Art. 16.
317 Ibid, Art. 17.
318 Ibid, Art. 21.
319 Supra, 2010 Guarani Aquifer Agreement, note 260, Art. 1.
320 Ibid, Art. 2.
321 Ibid, Art. 3.
322 Ibid.
323 Ibid, Art. 8.
324 Ibid, Arts. 9–11.
325 Ibid, Arts. 16–19.
326 Ibid, Art. 15.

yet entered into force and thus does not in itself represent customary international law.

Guidance from International Bodies

Over the past couple of decades, two U.N.-related bodies have provided guidance on the legal principles to apply to transboundary aquifers.

In 1992, the U.N. Economic Commission for Europe produced the Convention on the Protection and Use of Transboundary Watercourses and International Lakes ('UNECE Water Convention') that entered into force in 1996.[327] The scope of the UNECE Water Convention embraces both surface and ground waters that 'mark, cross or are located on boundaries between two or more States,' thus including aquifers, whether or not they are hydraulically connected to any surface water; however, the waters addressed in the convention stop at the shoreline of its state parties.[328] Parties to the UNECE Water Convention agree to 'take all appropriate measures to prevent, control and reduce any transboundary impact,'[329] and transboundary impact is defined to be 'any significant adverse effect on the environment... caused by a human activity.'[330] The convention lists three guiding principles: the precautionary principle, the polluter-pays principle and, in a reflection of the definition of sustainability that was given by the World Commission on Environment and Development in 1987,[331] a promise that 'the needs of the present generation are met without compromising the ability of future generations to meet their own needs.'[332] The Water Convention envisions a multidisciplinary approach to achieving its goals by mandating that the parties 'develop, adopt, implement and, as far as possible, render compatible relevant legal, administrative, economic, financial and technical measures....'[333] Twelve measures are listed, including pollution; reduction of nutrients and hazardous substances

327 Convention on the Protection and Use of Transboundary Watercourses and International Lakes, EIF 6 October 1996; UNTS 1936 (1992); ILM 1312 ('UNECE Water Convention'), *available at* http://www.unece.org/fileadmin/DAM/env/documents/2013/wat/ECE_MP.WAT_41.pdf (accessed 9 December 2015).

328 Ibid, Art. 1(1).

329 Ibid, Art. 2(1).

330 Ibid, Art. 1(2).

331 'Our Common Future,' *World Commission on Environment and Development* (1987), para. 27, *available at* http://www.un-documents.net/our-common-future.pdf (accessed 7 February 2016). 'Our Common Future' is also known as the Bruntland Report, after its Chair, Gro Harlem Brundtland.

332 Supra, UNECE Water Convention, note 327, Art. 2(5).

333 Ibid, Art. 3(1).

from industrial and municipal wastewater and from diffuse sources such as agriculture; application of environmental impact assessments; promotion of sustainable water-resources management, including the application of the ecosystems approach; development of contingency planning; and reduction of the risk of accidental pollution. One of the 12 measures specifically applies to groundwater, as parties are advised to 'take additional specific measures to prevent the pollution of groundwaters.'[334] Thus, transboundary aquifers benefit from all of the mandated measures that apply to surface waters, and then must also be protected by additional prophylactic but unspecified activities that would apply only to sub-surface water resources.

The UNECE Water Convention thus incorporates some standard tenets of international environmental law—the precautionary principle, the polluter-pays principle and the ecosystem approach—and the standard of preventing 'significant' impact is adopted. Some of the customary principles of water law that were discussed above also appear in the convention—monitoring,[335] information and data exchange,[336] joint bodies[337] and notification of 'any critical situation that may have transboundary impact.'[338] The UNECE Water Convention goes on to state that parties 'shall ensure' that information regarding transboundary waters is made available to the public, including the condition of the waters, current and planned measures for prevention, control and reduction of transboundary impact and the effectiveness of those measures.[339]

The Water Convention requires its parties to conclude cooperative agreements for the management of shared resources. In recognition of the unique character of transboundary aquifers and to provide guidance on the 'additional specific measures' to protect groundwater, the UNECE established the Core Group on Groundwater, led by Attila Tanzi and Alexandros Kolliopoulos, which produced the nine Model Provisions on Transboundary Groundwaters ('Model Provisions') that were adopted in 2012.[340] The foreword of the Model Provisions states that they 'reflect the current state of international water

334 Ibid, Art. 3(1)(k).

335 Ibid, Art. 4, 11.

336 Ibid, Art. 6, 13.

337 Ibid, Art. 9(2).

338 Ibid, Art. 14.

339 Ibid, Art. 16.

340 'United Nations Economic Commission for Europe Model Provisions on Transboundary Groundwaters,' *United Nations* (2014), *available at* https://www.unece.org/fileadmin/DAM/env/water/publications/WAT_model_provisions/ece_mp.wat_40_eng.pdf (accessed 10 February 2016).

law with regard to transboundary groundwaters.'[341] In keeping with generally accepted tenets of customary international water law, the provisions encourage parties to avoid transboundary impact,[342] utilize groundwater in an equitable and reasonable manner,[343] exercise the precautionary principle to protect aquifers[344] and exchange information and data.[345] In addition, the Model Provisions recommend integrated management of surface water and groundwater,[346] joint management plans,[347] use of environmental impact assessments where a planned activity is 'likely to have a significant effect on transboundary groundwaters,'[348] and establishment of a joint body to implement and coordinate groundwater management.[349] The Model Provisions drew inspiration from prior sets of guidelines, including the more detailed 1989 UNECE Charter on Groundwater Management.[350]

Five years after adoption of the UNECE Water Convention, the U.N. Convention on the Law of the Non-navigational Uses of International Watercourses ('U.N. Watercourses Convention') was adopted by the U.N. General Assembly and opened for signature in 1997.[351] While its provisions would not seem at first glance to be controversial, the UNILC required 27 years to produce the text of U.N. Watercourses Convention,[352] and nearly 20 years passed until the requisite 35 parties ratified it to cause its entry into force in 2014. Like the UNECE Water Convention, the U.N. Watercourses Convention focuses on protection of surface water but also includes groundwater within its

341 Ibid, p. iii.

342 Ibid, p. 5, Provision 1(1).

343 Ibid, Provision 1(2).

344 Ibid, p. 9, Provision 5.

345 Ibid, p. 12, Provision 6.

346 Ibid, p. 9, Provision 4.

347 Ibid, p. 12, Provision 7.

348 Ibid, p. 13, Provision 8.

349 Ibid, p. 14, Provision 9.

350 'Economic Commission for Europe Charter on Groundwater Management,' *United Nations* (1989), *available at* http://faolex.fao.org/docs/pdf/eur16399.pdf (accessed 10 February 2016).

351 Convention on the Law of the Non-Navigational Uses of International Watercourses, 21 May 1997, ILM 36 ('U.N. Watercourses Convention'), *available at* http://legal.un.org/ilc/texts/instruments/english/conventions/8_3_1997.pdf (accessed 9 December 2015).

352 The U.N. initiated the project that resulted in the U.N. Watercourses Convention in 1970. *See* 'Progressive Development and Codification of the Rules of International Law Relating to International Watercourses,' United Nations General Assembly Resolution 2669 (XXV) (8 December 1970), *available at* http://daccess-dds-ny.un.org/doc/RESOLUTION/GEN/NR0/349/34/IMG/NR034934.pdf?OpenElement (accessed 10 December 2016).

scope. However, unlike the UNECE Water Convention, the U.N. Watercourses Convention does not include aquifers that are not connected to a surface water system. Instead, by limiting its jurisdiction to 'a system of surface waters and groundwaters constituting by virtue of their physical relationship a unitary whole and normally flowing into a common terminus,' the U.N. Watercourses Convention takes a river basin approach and covers only waters, including groundwaters, that are connected to surface water.[353] Fossil aquifers therefore do not fall within its ambit, and aquifers that are only connected to other aquifers that are not themselves connected to surface waters would also fall outside the scope of the U.N. Watercourses Convention. Applicability of the U.N. Watercourses Convention to transboundary offshore aquifers would depend on the particular circumstances of the aquifer. The convention may very well apply to offshore aquifers that are connected to onshore aquifers that are then connected to surface waters. Unless there is a connection to surface waters, though, transboundary offshore aquifers would not be governed by the U.N. Watercourses Convention. Given these limitations, application of the U.N. Watercourses Convention to offshore aquifers is questionable.

With respect to groundwaters, familiar principles once again appear and apply both to surface waters and groundwaters: reasonable, equitable, optimal and sustainable utilization[354] through consideration of various factors;[355] a duty to cooperate in developing and protecting the watercourse;[356] prevention, elimination and/or mitigation of significant harm by 'all appropriate measures';[357] cooperation 'on the basis of sovereign equality, territorial integrity, mutual benefit and good faith;'[358] regular exchange of available data and information;[359] and prior notice of 'planned measures which may have a significant adverse effect upon other watercourse States.'[360] States may consider establishing joint bodies, but are not required to do so.[361] Instead of the polluter-pays principle, the U.N. Watercourses Convention provides that parties shall simply 'discuss the question of compensation.'[362] In the event of conflict

353 Supra, UN Watercourses Convention, note 351, Art. 2(a).
354 Ibid, Art. 5(1).
355 Ibid, Art. 6.
356 Ibid, Art. 5(2).
357 Ibid, Art. 7(1).
358 Ibid, Art. 8(1).
359 Ibid, Art. 9.
360 Ibid, Art. 12.
361 Ibid, Art. 8(2).
362 Ibid, Art. 7(2).

over uses of the watercourse, 'requirements for vital human needs' are to be given 'special regard.'[363]

The U.N. Watercourses Convention sets forth less stringent requirements for protection of surface waters and groundwaters than are present in the UNECE Water Convention, while maintaining and advancing some of the critical principles of the customary international water law. Another key difference is that the U.N. Watercourses Convention does not apply to aquifers that are not hydraulically connected to an international watercourse, thus excluding fossil aquifers and other aquifers not receiving recharge from or discharging into surface water. In this way, the U.N. Watercourses Agreement also departs from the accepted hydrogeological definition of 'confined aquifer.'[364] Nevertheless, in the same year that the U.N. Watercourses Convention was adopted, the International Court of Justice in the Gabcíkovo-Nagymoros Project Case ('Hungary v. Slovakia') took note of the convention and stated that it evidenced development of the principle outlined in the court's River Oder opinion that a 'community of interest in a navigable river becomes the basis of a common legal right.'[365]

The U.N. Watercourses Convention contains a conflict-of-laws provision stating that 'nothing in the present Convention shall affect the rights or obligations of a watercourse State' under agreements entered into prior to becoming a party to the U.N. Watercourses Convention, and further provides that parties only have to 'consider' harmonizing prior agreements with the U.N.

363 Ibid, Art. 10(2).

364 'First Report on Shared Natural Resources, A/CN.4/533,' Yearbook of the International Law Commission, Vol. II(1), *United Nations International Law Commission*, pp. 117–138 (2003) ('First report on shared natural resources'), p. 125, para. 30, *available at* http://legal.un.org/docs/?path=../ilc/publications/yearbooks/english/ilc_2003_v2_p1.pdf&lang=EFSRA (accessed 5 March 2016). As acknowledged by Chusei Yamada, the Special Rapporteur for the U.N. Draft Aquifer Articles (defined infra at page 57), the UNILC's definition of 'confined aquifers' differs from the hydrogeological definition, which refers to confining layers that may or may not permit a connection to surface waters. Ibid. *See also* Gabriel Eckstein, 'A Hydrogeological Perspective of the Status of Ground Water Resources under the UN Watercourse Convention,' *Columbia Journal of Environmental Law*, Vol. 30, pp. 525–564 (2005), pp. 550–551 and note 102.

365 Gabcíkovo-Nagymoros Project Case ('Hungary v. Slovakia;), 1997 ICJ No. 92, 85, *available at* http://www.icj-cij.org/docket/files/92/7375.pdf (accessed 5 March 2016). In the same paragraph, the ICJ also affirmed a state's right to 'an equitable and reasonable share' of natural resources, thereby recognizing a principle tenet of customary international law. Ibid.

Watercourses Convention.[366] At the time of its adoption, the UNECE Water Convention was open only to those states that were members of the UNECE, which are primarily European states.[367] However, in 2003 the parties to the UNECE Water Convention amended it to allow any state that is a member of the United Nations to accede to its provisions,[368] and those amendments entered into force in February 2013.[369] Thus, members of the United Nations can adopt the UNECE Water Convention, the U.N. Watercourses Convention, or both. The UNECE Water Convention is designed to be the overriding agreement among the parties, because by adopting the convention, the parties agree to 'enter into bilateral or multilateral agreements ... or adapt existing ones' to 'eliminate the contradictions with the basic principles of this Convention.'[370] This clear statement of priority raises interesting issues with respect to the provisions of the U.N. Watercourses Convention for states that are parties to both.

The non-binding U.N. Draft Articles on the Law of Transboundary Aquifers ('U.N. Draft Aquifer Articles') produced by the UNILC[371] and adopted by the

366 Supra, U.N. Watercourses Convention, note 351, Art. 3(2).

367 A list of the States that are parties to the UNECE Water Convention can be found at https://treaties.un.org/Pages/ViewDetails.aspx?src=TREATY&mtdsg_no=XXVII-5&chapter=27&lang=en (accessed 9 December 2015).

368 Decision III/1, Amendment to the Water Convention, ECE/MP.WAT/14, 28 November 2003, *available at* http://www.unece.org/fileadmin/DAM/env/documents/2004/wat/ece.mp.wat.14.e.pdf (accessed 9 February 2016).

369 'About the UNECE Water Convention,' UNECE, http://www.unece.org/env/water/text/text.html (accessed 9 February 2016). In order for nations outside the UNECE to accede to the Water Convention, all the parties to the Convention as of 28 November 2003 must ratify the amendments. 'UNECE Water Convention Goes Global,' UNECE, http://www.unece.org/index.php?id=32154 (accessed 7 February 2016). For a list of convention parties that have adopted the amendments and the dates on which the amendments were adopted, see 'Amendments to Articles 25 and 26 of the Convention on the Protection and Use of Transboundary Watercourses and International Lakes,' United Nations Treaty Collection, https://treaties.un.org/Pages/ViewDetails.aspx?src=TREATY&mtdsg_no=XXVII-5-b&chapter=27&lang=en (accessed 7 February 2016). In January 2016, Iraq announced its intention to be the first country outside of the UNECE to accede to the Water Convention. 'Iraq Confirms Progress Towards Accession to the UNECE Water Convention in 2016,' UNECE, http://www.unece.org/info/media/news/environment/2016/iraq-confirms-progress-towards-accession-to-the-unece-water-convention-in-2016/doc.html (accessed 7 February 2016).

370 Supra, UNECE Water Convention, note 327, Art. 9(1).

371 'Shared Natural Resources,' Yearbook of the International Law Commission, Vol. II(2), *United Nations International Law Commission*, pp. 17–43 (2008) pp. 19–22, para. 53,

U.N. General Assembly[372] in 2008, enjoy an interesting genesis and a surprising connection to oil and gas. In spite of the strong endorsement of the Special Rapporteur to the U.N. Watercourses Convention,[373] aquifers not connected to surface waters were not included in the scope of that convention and thus were reserved for separate treatment. During subsequent consideration of a separate legal document focused on aquifers, some states opined that, while the U.N. Watercourses Convention addressed groundwater as part of a 'unitary whole' that included related surface waters,[374] aquifers not connected to surface waters were more similar to natural resources such as oil and gas[375] over which states held sovereignty.[376] Thus, shared resources such as oil, gas and aquifers not connected to surface waters were included together in the ILC's 2002 programme of work,[377] since all three of these resources shared similar characteristics of being subterranean, transboundary and nonrenewable. Furthermore, a distinction was made between international aquifers—in which an aquifer and a river are hydrologically linked with an international border intersecting one or the other—and transboundary aquifers, where the aquifer is transected by an international boundary.[378]

In very short order, controversy arose on two concepts. First, states objected to the term 'shared,' since acceptance of that term could limit the extent of

('Draft Aquifer Articles'), *available at* http://legal.un.org/docs/?path=../ilc/publications/yearbooks/english/ilc_2008_v2_p2.pdf&lang=EFSR (accessed 14 March 2016).

372 The Law of Transboundary Aquifers, G.A. Res. 63/124, U.N. Doc. A/RES/63/124 (Jan. 15, 2009) ('The Law of Transboundary Aquifers G.A. Resolution'), *available at* http://www.un.org/en/ga/search/view_doc.asp?symbol=A/RES/63/124 (accessed 23 December 2015).

373 'Second Report of the Special Rapporteur on the Law of the Non-navigational Uses of International Watercourses,' Yearbook of the International Law Commission, Vol. II(1), *United Nations International Law Commission*, pp. 113–128, (1994), p. 114, para. 2–5, *available at* http://legal.un.org/docs/?path=../ilc/publications/yearbooks/english/ilc_1994_v2_p1.pdf&lang=EFSRAC (accessed 6 March 2016).

374 Supra, 'First Report on Shared Natural Resources,' note 364, p. 121, para. 12.

375 Ibid, p. 122, para. 13.

376 'Shared Natural Resources: First Report of the Special Rapporteur (Concluded),' Yearbook of the International Law Commission, Vol. I, *United Nations International Law Commission*, pp. 196–203 (2003), ('Shared Natural Resources: First Report of the Special Rapporteur (Concluded)), p. 201, para. 46, *available at* http://legal.un.org/docs/?path=../ilc/publications/yearbooks/english/ilc_2003_v1.pdf&lang=EFSRA (accessed 14 March 2016).

377 Yearbook of the International Law Commission, Vol. II(2), *United Nations International Law Commission* (2002), p. 11, para. 20, and p. 100, para. 518 (a), *available at* http://legal.un.org/docs/?path=../ilc/publications/yearbooks/english/ilc_2002_v2_p2.pdf&lang=EFSRAC (accessed 14 March 2016).

378 Supra, 'First Report on Shared Natural Resources,' note 364, p. 126, para. 38.

'permanent sovereignty' over the natural resources.[379] Secondly, states felt that practices regarding oil and gas were so numerous and divergent that customary laws of oil and gas had not yet fully developed to the point where they could be described.[380] Therefore, the ILC determined that aquifers not connected to surface waters would become the first topic of focused study of shared natural resources, and a decision would be made at a later time about oil and gas.[381] At the same time, members of the ILC recognized that, due to their complexity and vulnerability, groundwaters warranted greater protection than surface waters and also observed that a purely legal approach to the preservation of groundwaters would not be adequate—a successful solution should involve a mix of social, political and economic measures informed by a deep understanding of the vulnerability of groundwaters.[382]

As a consequence of the mandate from states to preserve sovereignty over natural resources, the term 'aquifer' refers not only to the geological formation itself but also to the water therein.[383] Article 3 states that '[e]ach aquifer State has sovereignty over the portion of a transboundary aquifer or aquifer system located within its territory,' thus giving states full sovereignty over both the water-bearing geological formations and the water contained therein that lay within their borders. A transboundary aquifer is defined as an aquifer or aquifer system lying in two or more states, thus excluding international aquifers where the recharge zone is in one state and the related surface water is in another.[384] Commentators such as Professor Stephen McCaffrey, who served as one of the Special Rapporteurs for the U.N. Watercourses Convention, have objected to a grant of sovereignty over the water, arguing that such a characterization cannot be reconciled with the concept of a community of interest

379 Yearbook of the International Law Commission, Vol. I, *United Nations International Law Commission* (2003), p. 189, para. 3 ('2003 Yearbook of the International Law Commission, Vol. I'), *available at* http://legal.un.org/docs/?path=../ilc/publications/yearbooks/english/ilc_2003_v1.pdf&lang=EFSRA (accessed 14 March 2016).

380 Ibid.

381 Supra, 'First Report on Shared Natural Resources,' note 364, pp. 119–120, para. 4–5. Ultimately, the UNILC decided against trying to codify the law of transboundary oil and gas reserves. 'Recursos Naturales Compartidos,' Yearbook of the International Law Commission, Vol. II(1), *United Nations International Law Commission*, pp. 83–87 (2010), p. 86, para. 15–17, *available at* http://legal.un.org/docs/?path=../ilc/publications/yearbooks/spanish/ilc_2010_v2_p1.pdf&lang=S (in Spanish) (accessed 14 March 2016).

382 Supra, 'Shared Natural Resources: First Report of the Special Rapporteur (Concluded),' note 376, p. 191, para. 11.

383 Supra, Draft Aquifer Articles, note 371, art. 2(a).

384 Ibid, Art. 2(c).

in water resources.[385] Salman M.A. Salman goes further in stating that limited territorial sovereignty, in which every riparian state has the right to equitable utilization of the water resource and a duty not to harm other users,[386] is the cornerstone of current international water law.[387] Professor Gabriel Eckstein notes that the U.N. Draft Aquifer Articles oblige states to exercise sovereignty in accordance with the other provisions, which effectively negates claims of absolute sovereignty over groundwaters.[388] However, according to hydrogeologist Shammy Puri, who was a principal advisor to the ILC as it was designing the U.N. Draft Aquifer Articles on transboundary aquifers, an aquifer by definition is both the immobile rock and the mobile groundwater. In a transboundary aquifer, states cannot be expected to relinquish sovereignty over immobile rocks (with potentially high-value minerals), though their sovereignty may be limited over the naturally mobile groundwater as it passes across national boundaries. In Puri's opinion, given the complexity of the rock-water hydrochemical and hydrodynamic interaction, sovereignty that attaches to the rock may still accompany the water moving through the rock, albeit to a lesser degree.[389] The issue of sovereignty over aquifer waters remains controversial and unsettled in the community of aquifer specialists.

In other ways, the provisions of the U.N. Draft Aquifer Articles mirror the familiar principles found in international surface water law generally and in the U.N. Watercourses Convention specifically. As noted by McCaffrey, the scopes and provisions of the U.N. Draft Aquifer Articles and the U.N. Watercourses Convention with respect to groundwater have a good deal of commonality, except for aquifers not connected to surface waters.[390] Thus, states agree to utilize aquifers equitably and reasonably,[391] and factors to be used in determining equitable and reasonable utilization are provided.[392] States agree to 'take

385 Stephen C. McCaffrey, 'The International Law Commission Adopts Draft Articles On Transboundary Aquifers,' *American Journal of International Law*, Vol. 103, pp. 272–293 (2009), p. 288.

386 Salman M.A. Salman, 'The Helsinki Rules, the U.N. Watercourses Convention and the Berlin Rules: Perspectives on International Water Law,' *International Journal of Water Resources Development*, Vol. 23, No. 4, pp. 625–640 (2007), p. 627.

387 Ibid, p. 628.

388 Gabriel E. Eckstein, 'Commentary on the U.N. International Law Commission's Draft Articles on the Law of Transboundary Aquifers,' *Colorado Journal of International Environmental Law & Policy*, Vol. 18, pp. 537–610 (2007), p. 651.

389 Shammy Puri, private correspondence with author, 20 February 2016.

390 Supra, McCaffrey, note 385, p. 283.

391 Supra, U.N. Draft Aquifer Articles, note 371, Art. 4.

392 Ibid, Art. 5.

all appropriate measures' not to cause significant harm to other states when utilizing an aquifer,[393] to 'cooperate on the basis of sovereign equality, territorial integrity, sustainable development, mutual benefit and good faith,'[394] to exchange data and information,[395] to protect and preserve ecosystems,[396] to prevent and control pollution,[397] to monitor the aquifer,[398] to establish and implement aquifer management plans,[399] and to give one another prior notice of planned activities that might affect the aquifer and thereby have a 'significant adverse effect' on another state.[400] In one aquifer-specific provision, states agree to identify and 'take appropriate measures to prevent and minimize detrimental impacts' in recharge and discharge zones.[401]

In their preambles, both the Guarani Aquifer Agreement and the ITAS MOU take note of U.N. General Assembly Resolution 63/124, which commends the U.N. Draft Aquifer Articles to states and encourages them to form group arrangements for proper management of aquifers while 'taking' the U.N. articles 'into account'.[402]

In summary, while centuries of practice and decades of codification have produced fairly well-defined governance principles for transboundary surface water, nations with transboundary aquifers have been slow to embrace formally the entirety of those principles with respect to groundwater. At the moment, the only practices that have been consistently included in those ratified treaties addressing transboundary aquifers are the sharing of data and the establishment of joint commissions.

Part 5: Perspectives and Conclusion

Harvesting the water in offshore aquifers would require a complex series of operations that might include seismic modeling, exploratory drilling, construction of infrastructure for extraction and transmission, temporary storage, treatment, and transportation and distribution to the final user(s). Each stage of

393 Ibid, Art. 6.
394 Ibid, Art. 7.
395 Ibid, Art. 8.
396 Ibid, Art. 10.
397 Ibid, Art. 12.
398 Ibid, Art. 13.
399 Ibid, Art. 14.
400 Ibid, Art. 15.
401 Ibid, Art. 11.
402 Supra, The Law of Transboundary Aquifers G.A. Resolution, note 372.

the process requires coordination among various parties as well as significant capital investment in sophisticated machinery and parts. At the moment—aside from active trade in bottled water—water is not traded as a global commodity, as hydrocarbons are. In the future, the price of water may vary wildly depending on local and regional circumstances such as availability and drought conditions. Should capital investors be confident that an adequate margin can be attained for the development of offshore aquifers, or should states dedicate government funds to the extraction of offshore fresh water, then at least some regions of the world will probably be experiencing such severe water scarcity that any new sources of fresh water will be viewed as highly prized assets. Urgent agricultural, industrial and/or domestic need for water may then turn regional or global attention to accessing and withdrawing waters from reserves that may be copious and pristine. In such a future, competition for freshwater resources may be intense and even violent as states seek to preserve human and other life. Crafting acceptable, multi-lateral approaches to common water resources would be more critical, and probably more difficult, than ever.

Assuming that TBOAs exist in sufficient quantities to make exploitation a sensible endeavor, certain climatic consequences should be analyzed before any offshore aquifers are developed on a large scale. For example, a 2012 study showed that groundwater withdrawals contributed 0.035 (±0.009) mm yr–1 to sea level rise in 1900 and increased to 0.57 (±0.09) mm yr–1 by 2000, with projections for a further increase to 0.82 (±0.13) mm yr–1 by 2050.[403] In addition to contributing to sea level rise, sheltered groundwater brought to the surface is exposed to evaporation, and water vapor is by far the most potent greenhouse gas, constituting an estimated 60% of the global warming effect.[404] However, melting polar regions and glaciers will contribute far more to sea level rise than groundwater extractions,[405] and the vast expanses of the ocean provide much greater surfaces to engage in evaporative processes than does groundwater

403 Yoshihide Wada, Ludovicus P. H. van Beek, Frederiek C. Sperna Weiland, Benjamin F. Chao, Yun-Hao Wu, Marc F. P. Bierkens, 'Past and Future Contribution of Global Groundwater Depletion to Sea-level Rise,' *Geophysical Research Letters*, Vol. 39, Issue 9 (2012), http://onlinelibrary.wiley.com/doi/10.1029/2012GL051230/full (last accessed 28 December 2015).

404 'It's Water Vapor, Not the CO2', *American Chemical Society*, http://www.acs.org/content/acs/en/climatescience/climatesciencenarratives/its-water-vapor-not-the-co2.html (last accessed 28 December 2015). *See also*, 'Water Vapor Confirmed as Major Player in Climate Change,' *NASA*, (November 17, 2008), *available at* http://www.nasa.gov/topics/earth/features/vapor_warming.html.

405 'Intergovernmental Panel on Climate Change: Climate Change 2014 Synthesis Report,' *WMO, UNEP* (2015), pp. 13, 16, *available at* https://www.ipcc.ch/pdf/assessment-report/ar5/syr/SYR_AR5_FINAL_full_wcover.pdf (accessed 11 February 2016).

brought to the surface. Therefore, concerns about contribution to sea level rise and climate change should not prevent development of offshore aquifers.

In presenting guidance on developing a legal regime for TBOAs, the first four chapters of this paper each addressed a different topic. The first chapter presented scientific arguments explaining and supporting the presence of unknown quantities of fresh to brackish water in continental shelf aquifers around the globe. These aquifers come in two types: those unconnected to another source of fresh water recharge and those hydraulically connected to a land-based aquifer. Since all of the aquifers discovered so far have been found in either the territorial sea or the EEZ of one state, no transboundary issues have yet arisen. Should future events make utilization of these freshwater caches both technically and financially feasible, the inevitable discovery of TBOAs that are common to two or more countries would prompt questions about proper governance. At least three major legal regimes should be consulted for guidance. The application of UNCLOS, which purports to codify customs and practices of a maritime regime, was discussed in the second chapter. TBOAs share many characteristics with offshore hydrocarbon deposits, whose contribution to the energy needs of a growing economy rapidly spawned a powerful industry with a well-formed set of practical treaties and legal structures that are described in Chapter 3. Finally, since TBOAs contain fresh water and the principles of offshore hydrocarbon development took inspiration from the customary law regime that formed for land-based fresh water, global approaches to sharing both surface fresh water and groundwater were featured in Chapter 4.

As a threshold issue, sovereignty over the shared water resources in TBOAs will need to be discussed and agreed upon, and the U.N. has presented guidance through several resolutions. In its 1962 Resolution 1803 (XVII) entitled 'Permanent Sovereignty over Natural Resources,' the U.N. General Assembly ('UNGA') recognized the 'free and beneficial exercise of the sovereignty of peoples and nations over their natural resources,' while at the same time cautioning that such exercise 'must be furthered by the mutual respect of States based on their sovereign equality.'[406] With regard to commonly held natural resources, UNGA Resolution 3129 (XXVIII), passed in December 1973, advises states to cooperate on exploitation of those natural resources through

406 'Permanent Sovereignty over Natural Resources,' United Nations General Assembly Resolution 1803 (XVII) (1962), *available at* http://www.un.org/ga/search/view_doc.asp?symbol=A/RES/1803%28XVII%29 (accessed 10 February 2016).

information exchange and prior consultation.[407] The following year, the theme of cooperation in exploitation of shared natural resources through information exchange and prior consultation was reinforced in Chapter 3 of the Charter of Economic Rights and Duties of States,[408] while Chapter 2 of the Charter reaffirmed the concept of full sovereignty over natural resources.[409] Thus, the U.N. has been consistent in stating that states have sovereignty over their natural resources, but when those resources are held in common, the exercise of ownership must be tempered by respect for other states and characterized by cooperation in exploitation. Likewise, the ICJ has opined that navigable watercourses represent a community of interest,[410] which would result in limitations on full sovereignty over a shared natural resource.

Additional guidance on sovereignty can be found in UNCLOS. Although not all states are parties to UNCLOS, with the United States being one of the more notable non-signatories,[411] the provisions of UNCLOS represent a codification of customary practices that have developed between and among states. Both the 12-mile limit of the territorial sea and the 200-mile limit of the EEZ have been widely accepted in theory and in practice. Likewise, absolute sovereignty over resources in the territorial sea and sovereign rights over seabed resources in the EEZ have been uniformly recognized and accepted by states. Given the technical and practical challenges inherent in the development of seabed resources, any development of offshore aquifers will likely take place no further than 200 miles from shore, and thus within the sovereign jurisdiction of one or more states. Where an offshore aquifer straddles a boundary between two or more states, both UNCLOS and the ICJ encourage those states to collaborate to find an equitable solution to resource management and development. In order to determine what might represent an equitable solution, two systems

407 'Cooperation in the Field of the Environment Concerning Resources Shared by Two or More States,' United Nations General Assembly Resolution 3128 (XXVIII) (1973), *available at* http://daccess-dds-ny.un.org/doc/RESOLUTION/GEN/NR0/282/01/IMG/NR028201.pdf?OpenElement (accessed 31 December 2015).

408 'Charter of Economic Rights and Duties of States,' United Nations General Assembly Resolution 3281 (XXIX), Art. 3(1974), *available at* http://www.un-documents.net/a29r3281.htm (last accessed 31 December 2015).

409 'Charter of Economic Rights and Duties of States,' United Nations General Assembly Resolution 3281 (XXIX), Art. 2(1) (1974), *available at* http://www.un-documents.net/a29r3281.htm (last accessed 31 December 2015).

410 Supra, Gabcíkovo-Nagymoros Project Case (Hungary v. Slovakia), note 365.

411 'Brief History of U.S. Efforts Relating to the Law of the Sea,' *U.S. Department of State Diplomacy in Action*, http://www.state.gov/e/oes/lawofthesea/179798.htm (last accessed 29 December 2015).

of practice would serve well as analogous precedents—offshore hydrocarbon development and land-based surface and groundwater laws and practices.

Customs and practices for development of common offshore hydrocarbon fields have evolved quite rapidly since the first wells were drilled in the 1950s, due to market-driven demands for efficient and orderly extraction and sale of an increasingly indispensable global commodity. Following practices for land-based hydrocarbon deposits, the offshore development of common fields first adopted a unitization approach to extraction that involved the joint appointment of a single operator in order to maximize the recovery of fluids. Unitization then evolved into joint development, whereby states appoint a common developer to both seek and extract hidden hydrocarbons within a defined geographical area. Since states voluntarily enter into JDAs and negotiate commercial terms that are acceptable to both sides, the final form of JDAs should by definition represent an equitable outcome as required by UNCLOS and the ICJ. One could argue that the presence and power of hegemonic states could skew a JDA toward a less-than-equitable outcome, but the presence of a valuable, marketable commodity gives negotiating power to owners and co-owners of the commodity and could actually enhance the bargaining positions of weaker states. Securing a mutually acceptable approach to offshore development through a treaty and its ensuing commercial contracts minimizes conflict, provides certainty, enhances efficiency and reduces costs.

Offshore aquifers share many similarities with offshore hydrocarbons, in that both are fluid, both are finite, both are found in geological formations with confining layers and both can be fouled by improper exploration and/or exploitation techniques. The customary laws and practices for transboundary offshore hydrocarbons that encourage joint development to maximize recovery of hydrocarbons would also serve to maximize recovery of water in TBOAs.

As discussed in Chapter 3, land-based water law principles influenced the legal regimes that evolved for offshore oil and gas development. Two staples of surface water law have become acknowledged tenets of law for all shared resources: equitable and reasonable utilization of the resource and no significant harm to a neighbor that shares the resource. While there is a plethora of treaties embodying those tenets in the lexicon of surface waters, the law and practice of land-based groundwater resources have not yet reached a stage of maturity where they can serve as models or precedents. Indeed, current practice for land-based aquifers still seems to follow the law of capture in an unregulated race to extract the resource.[412]

412 Renee Martin-Nagle, 'Fossil Aquifers: A Common Heritage of Mankind,' *Journal of Energy and Environmental Law*, Vol. 2, pp. 39–60 (2011), pp. 40, 49–51.

Offshore aquifers resemble land-based aquifers in that both contain fresh to brackish water that would need little to no treatment prior to utilization. Some are freestanding with no connection to any other water resource, and some are connected to another aquifer which may be land-based. Those offshore aquifers that have no connection to land should follow governance regimes for other marine resources. With respect to offshore aquifers that are connected to a land-based source of recharge, an argument could be made that the legal regime for land-based aquifers should apply. However, as noted, the currently meager legal regime for land-based aquifers does not adequately preserve the resource nor does it serve to maximize recovery. Following land-based practices in this case would probably not result in the most complete or efficient exploitation of the resource. Since offshore aquifers without a connection to land would exist in the absence of any dependent human communities or ecosystems, full development of the resource could proceed without concern for impact on local populaces or endogenous species.

Hydrogeologist Todd Jarvis called for a unitization approach to land-based aquifers,[413] and similarly a joint development agreement in line with those used by the offshore hydrocarbon industry would seem to be the best legal regime for the development of transboundary offshore aquifers. One structure that is common to both hydrocarbon development and land-based aquifer development is a joint commission, and those commissions would doubtless be part of the JDAs negotiated between states for TBOAs. However, in this case the joint commissions should be given much wider management authority than they are for most land-based aquifers, where they currently serve only to exchange information.

Conclusion

Regardless of whether the legal regimes for offshore hydrocarbon development serve as models for development of offshore aquifers, states will have to consider what role the offshore hydrocarbon industry should play in exploiting offshore water reserves. Since the offshore hydrocarbon industry has decades of experience and practice in offshore exploration for and exploitation of liquid natural resources, considerations of efficiency would seem to dictate that states take advantage of the technological, and perhaps legal, expertise possessed by the industry. The establishment of another, wholly separate extractive industry focused on offshore resources does not seem to be a viable

413 W. Todd Jarvis, 'Unitization: A Lesson in Collective Action from the Oil Industry for Aquifer Governance,' *Water International*, Vol. 36:5, pp. 619–630 (2011), pp. 624–628.

possibility in the short term, and states will doubtless engage with the hydrocarbon industry in order to save time, effort and money. Since development of hydrocarbon reserves requires water, the hydrocarbon industry may request the right to utilize the water in offshore aquifers for their own purposes as a price for engaging in development of those aquifers, which may be a sensible solution.

Generally speaking, consumers of a water resource have more essential needs than the consumers of hydrocarbon, since water supports life for all land-based creatures, and hydrocarbons simply provide energy for humans. In a world where states and private industry opt for costly development of offshore water resources, water would undoubtedly have become a precious, vital commodity that may very well be priced accordingly. Moral and economic choices would have to be made about whether to benefit agricultural, industrial or domestic users, and whether parched ecosystems should be replenished. In a water-scarce world, proper solutions to these and other critical issues will require thoughtful dialogue and creative thinking.

In an earlier article, this author recommended that the water in fossil aquifers be viewed as a common heritage of mankind, to be shared with water-scarce states and also to be utilized to support ecosystems.[414] Since offshore aquifers contain water that has not yet been included in any state's water budget, perhaps a state that discovers offshore aquifers with reserves surplus to its needs might consider how to utilize the resource to benefit more than its own citizens. Further, perhaps states sharing a transboundary offshore aquifer might consider cooperating to benefit less fortunate states and parched ecosystems. With U.N. General Assembly resolutions and the ICJ pronouncing that sovereignty over natural resources is not unlimited, and states voluntarily ceding some sovereignty to enter into joint development arrangements, going a step further to spread the watery wealth in domestic and transboundary offshore aquifers with those less fortunate would exemplify the kind of unifying behavior that epitomizes the more evolved aspects of our human natures. Sharing the water with other species who share this planet with us would be even better.

414 Supra, Martin-Nagle, note 412, pp. 14, 59–60.

List of References

About the UNECE Water Convention, UNECE, http://www.unece.org/env/water/text/text.html (accessed 9 February 2016).

Agreement on the Continental Shelf Between Iceland and Jan Mayen, 22 October 1981, *available at* http://www.un.org/Depts/los/LEGISLATIONANDTREATIES/PDFFILES/TREATIES/ISL-NOR1981CS.PDF (accessed 3 December 2015).

Agreement between the Government of Australia and the Government of the Democratic Republic of Timor-Leste Relating to the Unitization of the Sunrise and Troubadour Fields, 6 March 2003, 2483 UNTS 317, *available at* http://www.un.org/Depts/los/LEGISLATIONANDTREATIES/PDFFILES/TREATIES/AUS-TLS2003UNI.PDF (accessed 3 December 2015).

Agreement between the Government of the United Kingdom of Great Britain and Northern Ireland and the Government of the Kingdom of Norway relating to the Exploitation of the Frigg Field Reservoir and the Transmission of' Gas therefrom to the United Kingdom, Cmnd 6491, (1976), *available at* https://treaties.un.org/doc/Publication/UNTS/Volume%201098/volume-1098-I-16878-English.pdf (accessed 28 November 2015).

Agreement between Japan and the Republic of Korea Concerning Joint Development of the Southern Part of the Continental Shelf Adjacent to the Two Countries, 30 January 1974, 1225 UNTS 114, *available at* https://treaties.un.org/doc/Publication/UNTS/Volume%201225/volume-1225-I-19778-English.pdf (accessed 3 December 2015).

Agreement between the Government of the Hashemite Kingdom of Jordan and the Government of the Kingdom of Saudi Arabia for the Management and Utilization of the Ground Waters in the Al-Sag/Al-Disi Layer, 30 April 2015, Saudi Arabia and Jordan ('Al Sag/Al-Disi Agreement'), *available at* http://internationalwaterlaw.org/documents/asia.html#Al-Sag/Al-Disi%20Aquifer (accessed 5 December 2015).

Agreement between the Libyan Arab Socialist People's Jamahariya and the Republic of Tunisia to Implement the Judgment of the International Court of Justice in the Tunisia/Libya Continental Shelf Case, 8 August 1988, *available at* http://www.un.org/Depts/los/LEGISLATIONANDTREATIES/PDFFILES/TREATIES/LBY-TUN1988CS.PDF (accessed 3 December 2015).

Agreement between Sudan and Saudi Arabia Relating to the Joint Exploitation of the Natural Resources of the Seabed and Subsoil of the Red Sea in the Common Zone, 16 May 1974, 952 UNTS 197.

Agreement on the Guarani Aquifer, 2 August 2010, Brazil, Argentina, Uruguay and Paraguay ('2010 Guarani Aquifer Agreement'), *available at* http://internationalwaterlaw.org/documents/s-america.html (accessed 5 December 2015).

Amendments to Articles 25 and 26 of the Convention on the Protection and Use of Transboundary Watercourses and International Lakes, United Nations Treaty Collection, https://treaties.un.org/Pages/ViewDetails.aspx?src=TREATY&mtdsg_no=XXVII-5-b&chapter=27&lang=en (accessed 7 February 2016).

Arrangement Relatif à la Protection, à l'Utilization et à la Réalimentation de la Nappe Souterraine Franco-suisse du Genevois, EIF 9 June 1978, The State Council of the Republic and Canton of Geneva and The Prefect of Haute-Savoie, *available at* http://www.internationalwaterlaw.org/documents/regionaldocs/franko-swiss-aquifer.html (accessed 6 December 2015).

Bakken, Tor Haakon; Mangset, Lars Erik: 'Sub-marine Groundwater for the Supply of Drinking Water. A Review of the Hydro-geological Potential and Its Technical and Economical Feasibility,' *Geophysical Research Abstracts* Vol. 12, EGU2010–3309 (2010).

Bakken, Tor Haakon; Ruden, Fridtjof & Mangset, Lars Erik: 'Submarine Groundwater: A New Concept for the Supply of Drinking Water,' *Water Resource Management*, Vol. 26, pp. 1015–1026 (2012).

Bastida, Ana E. et al: 'Cross-border Unitization and Joint Development Agreements: An International Law Perspective,' *Houston Journal of International Law*, Vol. 29, pp. 355–422 (2006–2007).

Boisson de Chazournes, Laurence: *Fresh Water in International Law* (Oxford University Press 2013).

Bourdet, J.; Kempton, R. & Michael, K.: 'Palaeo-formation Water Evolution in the Latrobe Aquifer, Gippsland Basin, South-eastern Australia Continental Shelf,' *Geofluids*, Vol. 14, pp. 1–24 (2014).

'Brief History of U.S. Efforts Relating to the Law of the Sea,' *U.S. Department of State Diplomacy in Action*, http://www.state.gov/e/oes/lawofthesea/179798.htm (accessed 29 December 2015).

Buzan, Barry, *Seabed Politics* (Praeger Publishers 1976).

Cameron, Peter D.: 'The Rules Of Engagement: Developing Cross-border Petroleum Deposits in the North Sea and the Caribbean,' *International & Comparative Law Quarterly*, Vol. 55, pp. 559–585 (2006).

Case Concerning the Continental Shelf (Libyan Arab Jarnahiriya/Malta), Judgment, I.C.J. Reports 1985, p. 13, *available at* http://www.icj-cij.org/docket/files/68/6415.pdf (accessed 20 November 2015).

Case Concerning the Continental Shelf (Tunisia/Libyan Arab Jamahiriya), Judgment, I.C.J. Reports 1982, p. 18, *available at* http://www.icj-cij.org/docket/files/63/6267.pdf (accessed 20 November 2015).

'Charter of Economic Rights and Duties of States,' United Nations General Assembly Resolution 3281 (XXIX), Art. 3(1974), *available at* http://www.un-documents.net/a29r3281.htm (accessed 31 December 2015).

Church, Thomas M.: 'An Underground Route for the Water Cycle,' *Nature*, Vol. 380, pp. 579–580 (1996).

Clark, Peter U. et al.: 'The Last Glacial Maximum,' *Science*, Vol. 325, pp. 710–714 (2009).

Clopton, Edwin L.: 'Ice as a Mineral,' *Rocks & Minerals*, Vol.69(2), pp. 90–103 (1994).

Cohen, Denis et al.: 'Origin and Extent of Fresh Paleowaters on the Atlantic Continental Shelf, USA,' *GROUND WATER*, Vol. 48(1), pp. 143–158 (2010)

Constitution of the Joint Authority for the Study and Development of the Nubian Sandstone Aquifer Waters, 1992, Libya, Chad, Sudan and Egypt, *available at* http://www-naweb.iaea.org/napc/ih/documents/Nubian/Nubian_final_MSP_Sandstone.pdf (Annex 7) (accessed 14 January 2016).

Convention on the Continental Shelf, 29 April 1958, 15 U.S.T. 471, T.I.A.S. No. 5578, 499 U.N.T.S. 311, *available at* http://cil.nus.edu.sg/rp/il/pdf/1958%20Convention%20on%20the%20Continental%20Shelf-pdf.pdf (accessed 14 January 2016).

Convention between the Government of the French Republic and the Government of the Spanish State on the Delimitation of the Continental Shelves of the Two States in the Bay of Biscay (Golfe de Gascogne/Golfo de Vizcaya), 29 January 1974, *available at* http://www.un.org/Depts/los/LEGISLATIONANDTREATIES/PDFFILES/TREATIES/FRA-ESP1974CS.PDF (accessed 3 December 2015).

Convention on the Law of the Non-navigational Uses of International Watercourses, 21 May 1997, ILM 36, *available at* http://legal.un.org/ilc/texts/instruments/english/conventions/8_3_1997.pdf (accessed 9 December 2015).

Convention on the Protection and Use of Transboundary Watercourses and International Lakes, EIF 6 October 1996; UNTS 1936 (1992); ILM 1312, *available at* http://www.unece.org/fileadmin/DAM/env/documents/2013/wat/ECE_MP.WAT_41.pdf (accessed 9 December 2015).

Convention relative à la protection, à l'utilisation, à la réalimentation et au suivi de la nappe souterraine franco-suisse du Genevois, EIF 1 January 2008, the Community of the 'Annemassienne' region, the Community of the 'Genevois' Rural Districts, and the Rural District of Viry and the Republic and Canton of Geneva, *available at* http://internationalwaterlaw.org/documents/europe.html#Franco-Swiss%20Genevese%20Aquifer (accessed 6 December 2015).

Custodio, E.; Edmunds, W.M.; & Travi, Y.: 'Management of Coastal Palaeowaters,' in W. M. Edmunds & C. J. Milne, eds., *Palaeowaters in Coastal Europe: Evolution of Groundwater since the Late Pleistocene*, pp. 313–327 (The Geological Society of London 2001).

Decision III/1, Amendment To The Water Convention, ECE/MP.WAT/14, 28 November 2003, *available at* http://www.unece.org/fileadmin/DAM/env/documents/2004/wat/ece.mp.wat.14.e.pdf (accessed 9 December 2015).

DeFoor, Whitney et al.: 'Ice Sheet-derived Submarine Groundwater Discharge on Greenland's Continental Shelf,' *Water Resources Research*, Vol. 47(W07549), pp. 1–14 (2011).

Eckstein, Gabriel E.: 'Commentary on the U.N. International Law Commission's Draft Articles on the Law of Transboundary Aquifers,' *Colorado Journal of International Environmental Law & Policy*, Vol. 18, pp. 537–610 (2007).

Eckstein, Gabriel & Eckstein, Yoram, 'A Hydrogeological Approach to Transboundary Ground Water Resources and International Law,' *American University International Law Review*, Vol. 19, pp. 201–258 (2003).

'Economic Commission for Europe Charter on Groundwater Management,' United Nations (1989), *available at* http://faolex.fao.org/docs/pdf/eur16399.pdf (accessed 10 February 2016).

'Economic Commission for Europe Model Provisions on Transboundary Groundwaters,' United Nations (2014), *available at* https://www.unece.org/fileadmin/DAM/env/water/publications/WAT_model_provisions/ece_mp.wat_40_eng.pdf (accessed 10 February 2016).

Edmunds, W. M. & Milne, C. J. eds.: Palaeowaters in Coastal Europe: Evolution of Groundwater since the Late Pleistocene (The Geological Society of London 2001).

Edmunds, W.M.: 'Palaeowaters in European Coastal Aquifers—the Goals and Main Conclusions of the PALAEAUX Project,' in W. M. Edmunds & C. J. Milne, eds., *Palaeowaters in Coastal Europe: Evolution of Groundwater since the Late Pleistocene*, pp. 1–16 (The Geological Society of London 2001).

Establishment of a Consultation Mechanism for the Northwestern Sahara Aquifer System (SASS), 2002, Algeria, Libya and Tunisia ('NWSAS Agreement'), *available at* http://www.fao.org/docrep/008/y5739e/y5739e05.htm (accessed 6 December 2015).

Fesharaki, Fereidun: 'Joint Development of Offshore Petroleum Resources: The Persian Gulf Experience?,' *Energy*, Vol. 6 (11), pp. 1325–1334 (1981).

Faure, Hugues; Walter, Robert C.; & Grant, Douglas R.: 'The Coastal Oasis: Ice Age Springs on Emerged Continental Shelves,' *Global and Planetary Change*, Vol. 33, pp. 47–56 (2002).

Framework Agreement between the Government of the United Kingdom of Great Britain and Northern Ireland and the Government of the Kingdom of Norway concerning Cross-Boundary Petroleum Co-operation, 4 April 2005, *available at* https://www.gov.uk/government/uploads/system/uploads/attachment_data/file/273282/6792.pdf (accessed 3 December 2015).

Gabcíkovo-Nagymoros Project Case (Hungary v. Slovakia), 1997 ICJ No. 92, *available at* http://www.icj-cij.org/docket/files/92/7375.pdf (accessed 23 December 2015).

GMR (Great Man-Made River) Water Supply Project, Libya, water-technology.net, http://www.water-technology.net/projects/gmr (accessed 6 December 2015).

Groen, J. et al.: 'Fresh and Moderately Brackish Groundwaters in Coastal Plains and Continental Shelves: Past and Ongoing Natural Processes,' *Proc. SWIM [Salt Water Intrusion Meeting] 16*, Poland, pp. 1–7 (2000).

Groen, J.; Velstrab, J. & Meesters, A.G.C.A.: 'Salinization Processes in Paleowaters in Coastal Sediments of Suriname: Evidence from ∂ ^{37}Cl Analysis and Diffusion Modelling,' *Journal of Hydrology*, Vol. 234, pp. 1–20 (2000).

Hathaway, John C. et al.: 'U.S. Geological Survey Core Drilling on the Atlantic Shelf,' *Science*, Vol. 206(4418), pp. 515–527 (1979).

Hinsby, K. et al.: 'The Ribe Formation in Western Denmark—Holocene and Pleistocene Groundwaters in a Coastal Miocene Sand Aquifer,' in W. M. Edmunds & C. J. Milne, eds., *Palaeowaters in Coastal Europe: Evolution of Groundwater since the Late Pleistocene*, pp. 29–48 (The Geological Society of London 2001).

Hoefel, F. G. & Evans, R. L.: 'Impact of Low Salinity Porewater on Seafloor Electromagnetic Data: A Means of Detecting Submarine Groundwater Discharge?,' *Estuarine, Coastal and Shelf Science*, Vol. 52, pp. 179–189 (2001).

Horigan, James E.: 'Unitization of Petroleum Reservoirs Extending Across Sub-Sea Boundary Lines of Bordering States in the North Sea,' *Natural Resources Law*, Vol. 7, pp. 67–76 (1974).

'How Much Water Is There On, In, and Above the Earth?,' The USGS Water Science School, http://water.usgs.gov/edu/earthhowmuch.html (accessed on 5 March 2016).

'Intergovernmental Panel on Climate Change: Climate Change 2014 Synthesis Report,' *WMO, UNEP* (2015), *available at* https://www.ipcc.ch/pdf/assessment-report/ar5/syr/SYR_AR5_FINAL_full_wcover.pdf (accessed 11 February 2016).

'Iraq Confirms Progress towards Accession to the UNECE Water Convention in 2016,' UNECE, http://www.unece.org/info/media/news/environment/2016/iraq-confirms-progress-towards-accession-to-the-unece-water-convention-in-2016/doc.html (accessed 7 February 2016).

'Is Water a Mineral?—Is Ice a Mineral?,' Geology.com, http://geology.com/articles/water-mineral/ (accessed 7 February 2016).

'It's Water Vapor, Not the CO2,' *American Chemical Society*, http://www.acs.org/content/acs/en/climatescience/climatesciencenarratives/its-water-vapor-not-the-co2.html (accessed 28 December 2015).

Jarvis, W. Todd, 'Unitization: A Lesson in Collective Action from the Oil Industry for Aquifer Governance,' *Water International*, Vol. 36:5, pp. 619–630 (2011).

Joint Declaration Of 27 September 1995 on Cooperation over Offshore Activities in the South West Atlantic, *available at* http://fiassociation.com/shopimages/pdfs/7.%20 1995%20Joint%20Declaration%20on%20Cooperation%20Over%20Offshore%20 Activities%20in%20the%20South%20West%20Atlantic..pdf (accessed 3 December 2015).

Kohout, F.A. et al.: 'Fresh Ground Water Stored in Aquifers under the Continental Shelf: Implications from a Deep Test, Nantucket Island, Massachusetts,' *Water Resources Bulletin*, Vol. 13(2), pp. 373–386 (1977).

Kooi H. & Groen, J.: 'Offshore Continuation of Coastal Groundwater Systems; Predictions Using Sharp-interface Approximations and Variable-density Flow Modelling,' *Journal of Hydrology*, Vol. 246, pp. 19–35 (2001).

Kriete, Cornelia; Suckow, Axel & Harazim, Bodo: 'Pleistocene Meteoric Pore Water in Dated Marine Sediment cores off Callao, Peru,' *Estuarine, Coastal and Shelf Science*, Vol. 59, pp. 499–510 (2004).

Kusano, Yukiko: 'Occurrence of Old Groundwater in a Volcanic Island on a Continental Shelf: An Example from Nakano-shima Island, Oki-Dozen, Japan,' *Journal of Hydrology*, Vol. 511, pp. 295–309 (2014).

Lagoni, Rainer: 'Oil and Gas Deposits Across National Frontiers,' *American Journal of International Law*, Vol. 73, pp. 215–243 (1979).

Lambeck, Kurt & Chappell, John: 'Sea Level Change through the Last Glacial Cycle,' *Science*, Vol. 292, pp. 679–686 (2001).

Management and Cooperation Agreement between the Government of the Republic of Senegal and the Government of the Republic of Guinea-Bissau, 14 October 1993, *available at* http://www.un.org/Depts/los/LEGISLATIONANDTREATIES/PDFFILES/TREATIES/SEN-GNB1993MC.PDF (accessed 3 December 2015).

Martin-Nagle, Renee, 'Fossil Aquifers: A Common Heritage of Mankind,' *Journal of Energy and Environmental Law*, Vol. 2, pp. 39–60 (2011).

Mao, Kangshan et al.: 'Distribution of living Cupressaceae Reflects the Breakup of Pangea,' *Proceedings of the National Academy of Science*, Vol. 109 (20), pp. 7793–7798 (2012).

Maritime Delimitation Treaty between Jamaica and the Republic of Colombia, 12 November 1993, Reg. No. I-30943, *available at* http://www.un.org/Depts/los/LEGISLATIONANDTREATIES/PDFFILES/TREATIES/JAM-COL1993MD.PDF (accessed 7 February 2016).

McCaffrey, Stephen C.: 'The International Law Commission Adopts Draft Articles on Transboundary Aquifers,' *American Journal of International Law*, Vol. 103, pp. 272–293 (2009).

McIntyre, Owen: 'The Role of Customary Rules and Principles of International Environmental Law in the Protection of Shared International Freshwater Resources,' *Natural Resources Journal*, Vol. 46, pp. 157–210 (2006).

Medium Sized Project Proposal, *Global Environment Facility* (2005), *available at* http://www-naweb.iaea.org/napc/ih/documents/Nubian/Nubian_final_MSP_Sandstone.pdf (accessed 6 December 2015).

Meisler, Harold; Leahy, P. Patrick & Knobel, Leroy L.: 'Effect of Eustatic Sea-Level Changes on Saltwater-Freshwater in the Northern Atlantic Coastal Plain,' Water-Supply Paper 2255, *U.S. Geological Survey*, pp. 1–28 (1984).

Memorandum of Understanding between Malaysia and the Kingdom of Thailand on the Establishment of the Joint Authority for the Exploitation of the Resources of the Seabed in a Defined Area of the Continental Shelf of the Two Countries in the Gulf of Thailand, 21 February 1979, *available at* http://cil.nus.edu.sg/rp/il/pdf/1979%20MOU%20between%20Malaysia%20and%20Thailand-pdf.pdf (accessed 3 December 2015).

Memorandum of Understanding between Malaysia and the Socialist Republic of Vietnam for the Exploration and Exploitation of Petroleum in a Defined Area of the Continental Shelf Involving the Two Countries, 05 June 1992, *available at* https://assets.documentcloud.org/documents/1352723/vtn-mal-mou-between-malaysia-and-vietnam-for-the.pdf (accessed 7 February 2016).

Memorandum of Understanding Relating to the Setting Up of a Consultative Mechanism for the Management of the Iullemeden Aquifer System, 20 June 2009, Mali, Niger and Nigeria ('2009 Illumeden MOU'), *available at* http://www.internationalwaterlaw.org/documents/regionaldocs/Iullemeden_MoU-2009.pdf (accessed 5 December 2015).

Memorandum of Understanding for the Establishment of a Consultation Mechanism for the Integrated Management of the Water Resources of the Iullemeden, Taoudeni/Tanezrouft Aquifer Systems (ITAS), 28 March 2014, Algeria, Benin, Burkina Faso, Mali, Mauritania, Niger and Nigeria, *available at* http://internationalwaterlaw.org/documents/africa.html#Nubian%20Sandstone%20Aquifer (accessed 5 December 2015).

Miyoshi, Masahiro: 'Some Comments on Legal Aspects of Precedents for Joint Development,' *Energy*, Vol. 6 (11), pp. 1359–1362 (1981).

Miyoshi, Masahiro: 'The Joint Development of Offshore Oil and Gas in Relation to Maritime Boundary Delimitation,' *International Boundaries Research Unit Maritime Briefing*, Vol. 2(5), pp. 1–51 (1999).

Moore, Willard S.: 'Large Groundwater Inputs to Coastal Waters Revealed by Ra enrichments,' *Nature*, Vol. 380, pp. 612–14 (1996).

Morgan, A.L.: 'The New Law of the Sea: Rethinking the Implications for Sovereign Jurisdiction and Freedom of Action,' *Ocean Development & International Law*, Vol. 27, pp. 5–29 (1996).

Murphy, J. Brendan; Nance, R. Damian & Cawood, Peter A.: 'Contrasting Modes of Supercontinent Formation and the Conundrum of Pangea,' *Gondwana Research*, Vol. 15, pp. 408–420 (2009).

North Sea Continental Shelf Cases (Federal Republic of Germany/Denmark; Federal Republic of Germany/Netherlands), Judgment, I.C.J. Reports, 1969, *available at* http://www.icj-cij.org/docket/files/52/5563.pdf (accessed 20 November 2015).

North Sea Continental Shelf Cases (Federal Republic of Germany/Denmark; Federal Republic of Germany/Netherlands), Separate Opinion of Justice Jessup, I.C.J. Reports, 1969, *available at* http://www.icj-cij.org/docket/files/51/5543.pdf (accessed 22 November 2015).

Observatory of the Sahara and Sahel is http://www.oss-online.org/en (accessed 23 December 2015).

Okeke, Chidimma C., 'Nigeria, Other African Countries Intensify Efforts on Groundwater Resources,' *Daily Trust*, 2 April 2014, *available at* http://www.dailytrust.com.ng/daily/index.php/environment/20541-nigeria-other-african-countries-intensify-efforts-on-groundwater-resources (accessed 7 February 2016).

Onorato, William T.: 'Apportionment of an International Common Petroleum Deposit,' *International & Comparative Law Quarterly*, Vol. 17, pp. 85–102, 87 (1968).

Ong, David M.: 'Joint Development of Common Offshore Oil and Gas Deposits: "Mere" State Practice or Customary International Law?,' *The American Journal of International Law*, Vol. 93(4), pp. 771–804 (1999).

Onorato, William T.: 'Apportionment of an International Common Petroleum Deposit,' *International & Comparative Law Quarterly*, Vol. 26, pp. 324–337 (1977).

Onorato, William T.: 'Joint Development of Seabed Hydrocarbon Resources: An Overview of Precedents in the North Sea,' *Energy*, Vol. 6 (11), pp. 1311–1324 (1981).

Onorato, William T.: 'A Case Study in Joint Development: The Saudi Arabia-Kuwait Partitioned Neutral Zone,' *Energy*, Vol. 10 (3/4), pp. 539–544 (1985).

'Our Common Future,' *World Commission on Environment and Development* (1987), *available at* http://www.un-documents.net/our-common-future.pdf (accessed 7 February 2016).

Penick, F.V.W.: 'The Legal Character of the Right to Explore and Exploit the Natural Resources of the Continental Shelf,' *San Diego Law Review*, Vol. 22, pp. 765–778 (1985).

'Permanent and Definitive Solution to International Problem of the Salinity of the Colorado River,' Minute 242 of the International Boundary & Water Commission, para. 5, 6, 30 August 1973, *available* at http://www.ibwc.gov/Files/Minutes/Min242.pdf (accessed 7 February 2016).

Person, Mark et al.: 'Pleistocene Hydrogeology of the Atlantic Continental Shelf, New England,' *GSA Bulletin*, Vol. 115 (11), pp. 1324–1343 (2003).

Person, Mark et al.: 'Pleistocene Hydrology of North America: The Role of Ice Sheets in Reorganizing Groundwater Flow Systems,' *Reviews of Geophysics*, Vol. 45 (Paper 2006RG000206), pp. 1–28 (2007).

Programme for the Development of a Regional Strategy for the Utilisation of the Nubian Sandstone Aquifer System (NSAS)—Terms of Reference for the Monitoring and Exchange of Groundwater Information of the Nubian Sandstone Aquifer System and Terms of Reference for Monitoring and Data Sharing, 5 October 2000, Libya, Chad, Sudan and Egypt, *available at* http://www.fao.org/docrep/008/y5739e/y5739e05.htm (accessed 5 December 2015).

Post, Vincent E.A. et al.: 'Offshore Fresh Groundwater Reserves as a Global Phenomenon,' *Nature*, Vol. 504, pp. 71–78 (2013).

Rogers, John J.W. & Santosh, M.: 'Supercontinents in Earth History,' *Gondwana Research*, Vol. 6 (3), pp. 357–369 (2003).

Salman, Salman M.A.: 'The Helsinki Rules, the U.N. Watercourses Convention and the Berlin Rules: Perspectives on International Water Law,' *International Journal of Water Resources Development*, Vol. 23, No. 4, pp. 625–640 (2007).

'Scientific Correspondence: Submarine groundwater discharge,' *Nature*, Vol. 382, pp. 121–122 (1996).

Shah, Tushaar: 'Groundwater Governance and Irrigated Agriculture,' Technical Paper No. 19, *Global Water Partnership* (2014) p. 8, *available at* http://www.gwp.org/Global/ToolBox/Publications/Background%20papers/GWP_TEC_19_web.pdf (accessed 5 December 2015).

Shiklomanov, Igor A.: 'World Fresh Water Resources,' *in* Peter H. Gleick, Editor, *Water in Crisis: A Guide to the World's Fresh Water Resources*, pp. 13–24 (Oxford University Press 1993).

Special Edition of TBA Map will be launched at WWF7 (7 April 2015), International Groundwater Resources Assessment Center, http://www.un-igrac.org/news/special-edition-tba-map-will-be-launched-wwf7 (accessed 5 December 2015).

Supplementary Agreement to the Treaty Between the Kingdom of the Netherlands and the Federal Republic of Germany Concerning Arrangements for Co-operation in the Ems Estuary, 8 April 1960 (Ems-Dollard Treaty), *United Nations Treaty Series*, Vol. 509, pp. 140–157 (1964), *available at* https://treaties.un.org/doc/Publication/UNTS/Volume%20509/v509.pdf (accessed 28 November 2015).

Szekely, Alberto: 'The International Law of Submarine Transboundary Hydrocarbon Resources: Legal Limits to Behavior and Experiences for the Gulf of Mexico,' *Natural Resources Journal*, Vol. 26, pp. 733–768 (1986).

Treaty between the Federal Republic of Nigeria and the Democratic Republic of São Tomé and Príncipe on the Joint Development of Petroleum and other Resources, in respect of Areas of the Exclusive Economic Zone of the Two States, 21 February 2001, *available at* http://www.un.org/Depts/los/LEGISLATIONANDTREATIES/PDFFILES/TREATIES/STP-NGA2001.PDF (accessed 3 December 2015).

Truman, Harry S.: 'Proclamation 2667—Policy of the United States with Respect to the Natural Resources of the Subsoil and Sea Bed of the Continental Shelf,' 28 September 1945. Online by Gerhard Peters and John T. Woolley, *The American Presidency Project*, http://www.presidency.ucsb.edu/ws/?pid=12332 (accessed 12 November 2015).

'UNECE Water Convention Goes Global,' UNECE, http://www.unece.org/index.php?id=32154 (accessed 7 February 2016).

The United Nations Convention on the Law of the Sea, 10 December 1982, 1833 U.N.T.S. 897, *available at* http://www.un.org/depts/los/convention_agreements/texts/unclos/unclos_e.pdf (accessed 3 December 2015).

Treaty relating to the utilization of waters of the Colorado and Tijuana Rivers and of the Rio Grande, 59. Stat 1219, T.S. 994, 9 Bevins 1166, 3 U.N.T.S. 313 (February 3, 1944), *available at* http://www.ibwc.state.gov/Files/1944Treaty.pdf (accessed 7 February 2016).

United Nations General Assembly Resolution 374 (IV) (1949), *available at* http://daccess-dds-ny.un.org/doc/RESOLUTION/GEN/NR0/051/93/IMG/NR005193.pdf?OpenElement (accessed 15 November 2015).

United Nations General Assembly Resolution 1105 (XI) (1957), *available at* http://legal.un.org/diplomaticconferences/lawofthesea-1958/docs/english/vol_II/5_RESOLUTION_Plenary_Mtgs_vol_II_e.pdf (accessed 15 November 2015).

United Nations General Assembly Resolution 1803 (XVII), 'Permanent Sovereignty over Natural Resources,' (1962), *available at* http://www.un.org/ga/search/view_doc.asp?symbol=A/RES/1803%28XVII%29 (accessed 10 February 2016).

United Nations General Assembly Resolution 3128 (XXVIII) (1973), 'Cooperation in the Field of the Environment Concerning Resources Shared by Two or More States,' *available at* http://daccess-dds-ny.un.org/doc/RESOLUTION/GEN/NR0/282/01/IMG/NR028201.pdf?OpenElement (accessed 31 December 2015).

United Nations General Assembly Resolution 2669 (XXV), 'Progressive Development and Codification of the Rules of International Law Relating to International Watercourses,' (1970), *available at* http://legal.un.org/ilc/documentation/english/a_cn4_244.pdf (accessed 7 February 2016).

United Nations General Assembly Resolution 63/124, U.N. Doc. A/RES/63/124, 'The Law of Transboundary Aquifers' (15 January 2009), *available at* http://www.un.org/en/ga/search/view_doc.asp?symbol=A/RES/63/124 (accessed 23 December 2015).

Varma, Sunil & Michael, Karsten: 'Impact of Multi-purpose Aquifer Utilisation on a Variable-density Groundwater Flow System in the Gippsland Basin, Australia,' *Hydrogeology Journal*, Vol. 20, pp. 119–134 (2011).

Wada, Yoshihide et al.: 'Past and Future Contribution of Global Groundwater Depletion to Sea-level Rise,' *Geophysical Research Letters*, Vol. 39, Issue 9 (2012), http://onlinelibrary.wiley.com/doi/10.1029/2012GL051230/full (accessed 28 December 2015).

'Water Vapor Confirmed as Major Player in Climate Change,' NASA, (November 17, 2008), *available at* http://www.nasa.gov/topics/earth/features/vapor_warming.html (accessed 7 February 2016).

'What is Groundwater?,' International Groundwater Resources Assessment Center, http://www.un-igrac.org/what-groundwater (accessed 5 December 2015).

Whiteman, Marjorie M.: 'Conference on the Law of the Sea: Convention on the Continental Shelf,' *American Journal of International Law*, Vol. 52, pp. 629–659 (1958).

Yearbook of the International Law Commission, Vol. I, 'Report to the General Assembly,' *United Nations International Law Commission*, p. 281, para. 19 (1949), *available at* http://legal.un.org/docs/?path=../ilc/documentation/english/reports/a_cn4_13.pdf&lang=E (last accessed 14 March 2016).

Yearbook of the International Law Commission, Vol. I, 'Regime of the High Seas: Report of Mr. François,' *United Nations International Law Commission*, pp. 179–187 (1950), *available at* http://legal.un.org/docs/?path=../ilc/publications/yearbooks/english/ilc_1950_v1.pdf&lang=E (last accessed 14 March 2016).

Yearbook of the International Law Commission, Vol. II, 'Regime of the High Seas: Rapport de J. P. A. François, Rapporteur special,' *United Nations International Law Commission*, pp. 36–52 (1950) (in French), *available at* http://legal.un.org/docs/?path=../ilc/publications/yearbooks/english/ilc_1950_v2.pdf&lang=E (accessed 14 March 2016).

Yearbook of the International Law Commission, Vol. II, 'Report of the International Law Commission to the General Assembly: Law of the Sea,' *United Nations International Law Commission*, pp. 254–301(1956) ('1956 UNILC Yearbook'), *available at* http://legal.un.org/docs/?path=../ilc/publications/yearbooks/english/ilc_1956_v2.pdf&lang=EFS (last accessed 14 March 2016).

Yearbook of the International Law Commission, Vol. II(1), 'Second Report of the Special Rapporteur on the Law of the Non-navigational Uses of International Watercourses,' *United Nations International Law Commission*, pp. 113–128, (1994), *available at* http://legal.un.org/docs/?path=../ilc/publications/yearbooks/english/ilc_1994_v2_p1.pdf&lang=EFSRAC (accessed 6 March 2016).

Yearbook of the International Law Commission, Vol. II(2), *United Nations International Law Commission* (2002), *available at* http://legal.un.org/docs/?path=../ilc/publications/yearbooks/english/ilc_2002_v2_p2.pdf&lang=EFSRAC (accessed 14 March 2016).

Yearbook of the International Law Commission, Vol. I, 'Shared Natural Resources: First Report of the Special Rapporteur (Concluded),' *United Nations International Law Commission*, pp. 196–203 (2003), *available at* http://legal.un.org/docs/?path=../ilc/publications/yearbooks/english/ilc_2003_v1.pdf&lang=EFSRA (accessed 14 March 2016).

Yearbook of the International Law Commission, Vol. II(1), 'First Report on Shared Natural Resources, A/CN.4/533,' *United Nations International Law Commission*, pp. 117–138 (2003) ('First report on shared natural resources'), *available at* http://legal.un.org/docs/?path=../ilc/publications/yearbooks/english/ilc_2003_v2_p1.pdf&lang=EFSRA (accessed 14 March 2016).

Yearbook of the International Law Commission, Vol. II(2), 'Shared Natural Resources,' *United Nations International Law Commission*, pp. 17–43 (2008) pp. 19–22, para. 53 ('Draft Aquifer Articles'), *available at* http://legal.un.org/docs/?path=../ilc/publications/yearbooks/english/ilc_2008_v2_p2.pdf&lang=EFSR (accessed 14 March 2016).

Yearbook of the International Law Commission, Vol. II(1), 'Recursos Naturales Compartidos,' *United Nations International Law Commission*, pp. 83–87 (2010), *available at* http://legal.un.org/docs/?path=../ilc/publications/yearbooks/spanish/ilc_2010_v2_p1.pdf&lang=S (in Spanish) (accessed 14 March 2016).

Zimmerman, Kim Ann: 'Pleistocene Epoch: Facts about the Last Ice Age,' livescience, http://www.livescience.com/40311-pleistocene-epoch.html (accessed 26 December 2015).

www.ingramcontent.com/pod-product-compliance
Lightning Source LLC
Chambersburg PA
CBHW071358230926
51846CB00002B/586
* 9 7 8 9 0 0 4 3 2 6 3 5 4 *